THE NORTH AMERICAN

MARIA THUN
BIODYNAMIC
CALENDAR
2017

CREATED BY
MARIA AND MATTHIAS THUN

Floris
Books

Compiled by Matthias Thun
Translated by Bernard Jarman
Additional astronomical material
by Wolfgang Held and Christian Maclean

Published in German under the title *Aussaattage*
English edition published by Floris Books

British Library CIP Data available

ISBN 978-178250-332-3
ISSN: 2052-577X

Printed in Poland

Contents

Walter Thun, Idyllic Spring, *oil, 48 x 69 cm, 1980*

Walter Thun's *Idyllic Spring*

If we spend some time thinking about this picture, we might feel like changing its title. The spring is only a small part of it – indeed, the central spring itself is almost hidden – and there is such manifold life here that we can almost feel ourselves transported to an early time of creation.

The glorious plant world covers the bare rocks, there's a frog which is closely tied to the watery element, and a lizard and curious snake to round off the image. Looking into the almost invisible cavern of the spring we can see a snail that lives entirely in water, and what appears to be a stone, but could also be a shell.

The entire cosmos is also incorporated into this painting. All kinds of stars can be found and in the background, the Moon so intimately linked to the movement of water and life, is patiently waiting.

Introduction

The emphasis in this year's calendar is on astronomical questions and how they relate to plant growth.

During an *opposition,* forces stream into the plant world from the constellations in which the two opposing luminaries are standing. Since constellations of a similar type are never opposite one another, a definitive effect of an opposition cannot be determined. If, for instance, one planet is in the Flower constellation of Libra, and another in Aries which is a Warmth-Fruit constellation, there is a mixed effect. It might be possible to identify one or the other influence after the event, but to predict with certainty is almost impossible.

With *trines* it is quite a different matter: their effects can often be seen in the way plants grow over time. Pages 25–27 give more details of this.

A third kind of effect comes about through *conjunctions* – when one luminary meets or passes by another. There is almost no positive influence on plant growth from conjunctions. Their effect becomes stronger if they are so close that the Moon or Sun covers a planet as seen from the Earth; astronomers call this an *occultation.* (In theory a planet can occlude another, but this is extremely rare, the last one was in 1818, the next in 2065). An occultation by the Moon can be observed, but one by the Sun cannot be seen. Sceptics often claim that these occultations cannot have any effect. When sowing trials are carried out during the period of an occultation, however, the missing influence of the occluded planet can subsequently be observed. It is as if the usual influence of the planet is 'blocked' when the Moon or Sun moves in front of it.

Another common misconception is that an occultation (or any other aspect of planets) occurs for a very limited period, such that a calculation given for Britain in GMT, for example, cannot have any effect in other parts of the world. This is a misunderstanding. All the aspects given are valid across the whole world, and a conjunction shown as taking place at 10 am EST takes place at the same time in London or in Moscow (though it will be 3 pm local time in London and 7 pm in Moscow).

Some years ago the International Astronomical Union demoted Pluto to 'dwarf planet', as it was deemed to be too small to be a real planet. However, plants seem to be unaware of this reclassification and continue to respond to its influences, which are no less significant than those of the larger planets. All the planets bring their own particular influence to bear on plant growth and on the weather. This is most noticeable when their influence has been blocked by an occultation, for example causing a deformation of the fruit.

For those interested in following the specific aspects of the planets during 2017, we are listing them in terms of number:
– 18 occultations (marked ☾)
– 14 planetary oppositions of Water-Earth constellations
– 18 planetary oppositions of Light-Warmth constellations
Additionally there are
– 9 Light (Flower) trines (marked △)
– 12 Water (Leaf) trines (marked △)
– 8 Warmth (Fruit) trines (marked △)
– 6 Earth (Root) trines (marked △)
The large number of trines in 2017 have a marked effect on plant growth, and our calendar shows how to make use of these influences.

Review of 2014 to 2016 and the Year Ahead

In our part of Germany, 2014 was a textbook year. An extended and uninterrupted spring, followed by a good summer and autumn, ensured a good harvest for farmers and gardeners. Bees also gathered so much nectar from flowers right through into the autumn that we had the highest yield of honey for many years. While winter did not see real cold or significant snowfalls, it was still clearly wintry.

The following year, 2015, started in a similar way but did not progress as we might have hoped. And 2016 has been really difficult. Spring started well, but beekeepers who had harvested honey during the spring flush had to return some. Although there was a reasonable honey flow in the summer, the continual change from hot to cooler weather meant that the bees had to use a lot of their stored honey. While in 2015 there was plenty of blossom honey to see the bees through the winter, this will probably not be the case in 2016. They will need to be given sugar together with the herb teas we recommend.

The first part if 2016 was also difficult in the garden and on the fields. Cool and watery influences were so strong that they affected hay and grain harvests.

Looking at 2017, the planets show an ambiguous pattern. In the first half of the year, influences from the constellation of Leo (which brings Warmth and Fruit influence) are missing. They only appear in the second half of the year to help the fruit and seed harvest in the northern hemisphere. In the southern hemisphere the Leo influences are less significant, and we need to look to the constellations of Sagittarius and Aries for that influence in spring and autumn.

We feel that at least the first half of 2017 will be dominated by cold and damp.

A History of Biodynamics

In recent years there has been a growing interest in lunar rhythms, so we shall take a brief look at the history of biodynamics in order to understand how this has come about.

For centuries, old farming wisdom was passed down, sometimes with little real understanding, but sufficient for subsistence living. Then in the nineteenth century the chemist Justus von Liebig (1803–73) changed the direction of agriculture. His research into plant nutrients led him to add certain elements – particularly nitrogen – to the soil, which led to significant increases in yields.

This new approach was widely embraced, but also meant a departure from the accepted principle of using compost and manure to improve soil fertility. Instead, manuring was seen as a way of feeding plants. Over time this led to a decline in soil quality and sometimes serious deficiency symptoms. Liebig recognised the problem, and in his 1844 *Chemische Briefe* (Letters on Chemistry) urged farmers not to ignore soil fertility. His plea was largely ignored, however, and today he's known as the 'father of the fertiliser industry'.

By 1924, soil conditions had degraded so much that a number of farmers approached Rudolf Steiner, the founder of anthroposophy, for help. The health and quality of their crops and livestock had deteriorated as a result of using chemical fertiliser. Steiner's subsequent Agriculture Course addressed how to

Mulching green manure

bring vitality to soil, and became the basis for biodynamic agriculture. A key principle is that fertilising means bringing life to the soil – to the earthy constituents of the soil.

From small beginnings, biodynamic agriculture spread rapidly after the Second World War. It has become an important ingredient in European agriculture and is now practiced in at least sixty countries around the world. The popularity of organic methods also increased dramatically in that time. But Steiner's biodynamic recommendations are not so easily understood; he repeatedly emphasised the importance of taking local conditions – including soil, climate, vegetation, animals and people – into account. It's not a one-size-fits-all approach and it requires an individual connection to the farm or garden.

Maria Thun's work

In the 1940s Maria Thun, who had grown up on a farm, came across the Agriculture Course. Much of its content reminded her of her father's old methods, and she resolved to test these new ideas. She was particularly struck by the idea that the growth of plants was related to the movement of stars and planets. She had heard of this from her parents, and the idea was self-evident to her. However, her parents' methods seemed rather imprecise, so she began systematic planting trials in 1952 and over the years published her results. After other practitioners had assessed her trials and corroborated her results, she felt confident enough to publish an annual sowing calendar from 1962.

Compost trials

Full Moon   First visible crescent after New Moon

It is well known that our planting trials have taken place, and will continue to take place, in biodynamic soil. In the late 1960s and 1970s Maria Thun was asked whether her results would also hold true in conventionally treated soil. Subsequent trials in our own fields (which had recently expanded) alongside conventional trials at agricultural colleges found that, provided certain basic agricultural rules were observed, the effects of the stars hold true for *all* plant growth.

In these trials we were of course reminded of Liebig's plea to farmers that particular attention be paid to the humus content of the soil and we saw to it that special consideration was given to the building up of humus. If a lot of fresh manure or plant material such as from green manure is applied directly to the soil, a high raw humus content will be achieved which is only partly available to plants. In order for humus to be biologically active it needs to either be composted or carefully worked into the soil in such a way that microorganisms are activated (see Further Reading, p.64, for detailed sources of information on the different forms of composting). To encourage microorganisms, earthworms and other soil fauna, muck should be spread on the green manure before turning it in. It should then be left to wither and dry out for a day or two and then turned lightly into the top soil. Once the green manure has been incorporated, three applications of barrel preparation can be given to support its transformation and to build soil humus. After a week or two it may then be ploughed in deeper. If the green manure isn't allowed to dry before being ploughed in, soil organisms may not be able to cope. In heavy soils it could then be many months before it can be processed by soil organisms, making good humus formation very hard to achieve. Done correctly, though, this approach can improve almost any soil, regardless of subsequent farming systems, and make it accessible to the influences of lunar constellations.

We found in our composting trials that the soil needs to contain at least 1.1% available humus for planetary and lunar influences to take effect. If the humus content is lower, these effects do not show. However, it is not only the humus content but also how water is dealt with, that determines how effective the lunar constellations will be. Indeed, most people know that the Moon is linked with tidal rhythms and not only is there a (half) daily rhythm of high and low tides, but there is monthly rhythm of particularly high tides (spring tides) around New Moon and Full Moon, and less extreme tides (neap tides) around the Half Moons.

What is special about this calendar?

The growing interest in lunar rhythms has seen many different lunar calendars spring up, and the *Maria Thun Biodynamic Calendar* is often lumped in with them. However, we don't consider this to be simply a lunar calendar because we look at both lunar *and* planetary influences. It may help new readers to explore this in a bit more detail; we hope that long-standing readers will bear with us.

Let's first of all revisit the effects of the Moon itself. Our earliest trials showed that the moon operates in different areas. One of these has to do with the watery element and is connected to the *phase* of the moon (Full Moon, waxing, waning, etc.). The Moon also acts as mediator for the *zodiac constellation* in which it is positioned. Here the phase is of only minor importance. The moon's phases are of special interest to the observer since they can be followed in the sky. To the soil and plants however their importance is minimal.

We might imagine that constellation effects are continually being mediated by the Moon, and that plants are therefore continually under their influence. Although this is partly true, plants can only fully respond to and process these

Heavy sprinkling, if used sensibly, should take place at night

Broad beans

influences if the soil in which they are growing has simultaneously been moved. This soil movement activates the microorganisms and they in turn give the plants a new orientation that enables them to respond to the current zodiacal influence. This fine and seemingly subtle action of the microorganisms, however, only occurs if the humus in the soil is in the right condition for the plants and is accessible to them.

The zodiacal influence of the Moon can be suppressed when too much fertiliser is applied (of either organic or inorganic origin) or by excessive irrigation. The soil organisms are then overwhelmed and only the moon's phases have any influence on the plants.

Let's now consider the effects of the planets. They also influence plant growth and weather, separately from the Moon. Seen from the Earth, they too wander through the zodiac and communicate what lives in the different constellations. In this case, however, it is coupled with the planet's own influence. If for example Neptune, which has a watery nature, moves through a Water constellation such as Cancer, its watery influence is magnified. Similarly if Venus, which carries light characteristics, passes through the Light constellation of Gemini, there will be a strong Light and Flower influence.

We can see therefore, it is not only the Moon that rules and mediates what occurs in the heavens but that all the planets work harmoniously together making it possible for human beings to intervene constructively in the process.

We should also briefly mention the effects of the Sun. Seen from the Earth, the Sun is also a wandering luminary. It gives us, of course, our two fundamental rhythms of day and night, and the seasons of the year. But it can also be thought of as a grand master working as the conductor of our cosmos. Steiner suggested that the Sun's character changes according to the constellation in which it stands. It would then be called a Gemini Sun, a Cancer Sun, and so on. And here's one quick example of the Sun's effects: beans such as broad beans are often attacked by aphids when there is a change in weather. By planting broad beans at Fruit times when the Sun is in the constellation of Aquarius, aphid attacks can be reduced.

A major difference between the *Maria Thun Biodynamic Calendar* and other lunar calendars is that others usually work with regular astrological *signs,* whereas this calendar uses the visible astronomical *constellations.* (The relationship between the signs and constellations is shown on p. 24.) This is significant because the visible constellations vary in size, and are not in the same positions as the signs. For that reason the dates given in this calendar are different from those of the fixed-length astrological signs. The importance of using the visible constellations for sowing and planting times has been confirmed in numerous trials over several decades.

For the astronomical data shown in this calendar we are of course dependent on various astronomical ephemerides, and each year the position of Sun, Moon and planets in relation to the constellations of the zodiac used here has to be calculated anew. For the English editions, the Central European Time given in the German calendar is converted to GMT or Eastern Standard/Daylight Time (for the North American version).

The Effect of the Sun on Plant Growth

Maria Thun

This and the following two articles were written some years ago but they relate directly to the theme of this calendar.

The Sun gives the Earth its two main rhythms of day and night, and of the seasons of the year. As well as showing which constellation the Moon is in, we also show the date when the Sun moves from one zodiac constellation to another. (Remember that these dates relate to the visible astronomical constellations, not to the astrological signs; for more information, see p. 24.)

Carrots

It's well known that the Earth rotates around its axis once every 24 hours, and that the stars of the zodiac can therefore be seen from any point on Earth, assuming it's not the daytime – when the Sun outshines them – or a cloudy night. If we look at the stars on a clear night at intervals of one or two hours, we can become aware of how they move steadily forwards. This east to west movement reflects the Earth's rotation and in 24 hours completes a full 360° circuit.

Rapeseed

Sunflower

False flax   *Linseed*

There is also a slower counter-movement of about 1° per day because of the Earth's annual revolution around the Sun.

Although the stars are not visible during the day, it is possible to calculate exactly which constellation is behind the Sun. At his observatory on the island of Hven, the great astronomer Tycho Brahe had a deep shaft sunk into the earth with a tall tower built above, so that he was able to observe a tiny portion of the sky during daytime, because all the sunlight was cut out. This greatly aided his astronomical observations.

The effect of the Sun on soil, and therefore plant growth, shouldn't be underestimated. In early editions of this calendar we presented results from trials that showed how soil composition subtly changes when the Sun moves from one constellation to another, because of a change in the activity of microorganisms in the soil. Soil analysis can give some indication of these changes.

Research by the microbiologist Erhard Ahrens has shown that nitrogen fixing bacteria (azatobacter) are especially active in the soil when the Sun is in particular constellations. Nitrogen fixation reaches its peak in May and September in the northern hemisphere, and in January and February in the southern hemisphere. This clearly indicates that nitrogen fixation is stimulated when the Sun is in the constellations of Taurus (May 14 – June 21) and Virgo (Sep 16 – Nov 1) and Capricorn (Jan 20 – Feb 15). The bacteria appear to rest during the times in between. Field trials and laboratory experiments have shown that when the Moon passes through these same constellations (two days in Capricorn, two days in Taurus, four days in Virgo) both the decomposing bacteria and nitrogen fixing azotobacter are significantly more active than at other times. It is as if the Moon's path through the zodiac and its effect is a kind of reflection of the Sun's.

The effect of the Sun in different constellations can also be found in cultivated plants. It was well known, for instance, to traditional farmers that oats and

Traditional grains for seed production

field beans sown in the second half of February or the first half of March – when the Sun is in the constellation of Aquarius – produce the most healthy plants and best yields. Beans sown later, when the Sun is in Pisces, are likely to attract aphids, while oats are more susceptible to fungal infections.

Carrots are similarly affected by the Sun's position in the zodiac. If they are sown when the Sun is in Pisces, their leaves grow rampant and the crowns have a tendency to turn green, both of which result in a poor flavour. If they are sown when the Sun is in Aries, they barely grow and are soon overrun by weeds. However, if they're sown when the Sun is in Taurus (April – May), early growth is rapid and the carrots are able to ripen in autumn. The Virgo Sun stimulates sugar formation and ripens the protein so that the carrots have a low nitrate content.

Plants grown for their oil such as rapeseed and sunflowers are also influenced by the Sun's position in the zodiac. Winter rape thrives best when sown with the Sun in Leo, spring sown oil crops are best sown when the Sun is in Aries (March – April) as this is a Warmth constellation. For their further cultivation and to encourage the productaion of oil, times when the Moon is in Light constellations should be chosen.

As a final word, it's interesting to note that earthworms are influenced by cosmic rhythms as well. This can be observed by counting the number of worms present at different soil depths. When the Moon is in Aries, Leo and Sagittarius, the majority of worms are at a depth of 100–120 cm (40–48 in); when the

Moon is in Taurus, Virgo or Capricorn, they're at 5–20 cm (2–8 in); and when the Moon is in Pisces, Cancer or Scorpio, they rise up to about 5 cm (2 in), even coming out of the ground overnight to wriggle on paths.

Maximising Plant Fertility and Nutritional Value

Maria Thun

The best nutrition is accomplished by making our farms and gardens receptive to the forces of the zodiac. Rudolf Steiner stated that 'the forces of the earth of the cosmos take effect through earthly substances' and so in this article we shall explore how that works in practice.

Cultivated plants have a capacity for producing nutritive substances including proteins, fats and carbohydrates. They also have a natural ability to reproduce, whether through seeds, or asexually like strawberries and potatoes. But there has been a gradual decline in fertility since the start of the twentieth century, which was one of the reasons that farmers asked for Rudolf Steiner's help and led to his Agriculture Course. Steiner stated that a plant's reproductive power comes from the inferior planets (Moon, Mercury, Venus) but that its nutritional qualities come from the superior plants (Mars, Jupiter, Saturn).

In grain crops, the two aspects are of course intimately linked in seed formation. Steiner indicated that regenerative power is stimulated by sowing crops 'close to winter', whereas sowing further away from winter improves their nutritional qualities. Sowing close to winter does not necessarily mean during the Christmas period as is often suggested. The old farmers referred to late sowings as 'Advent grains'. The constellation best suited to grain sowing is Sagittarius. Nowadays the Sun enters in Sagittarius around December 19 (which still leaves five days of Advent for sowing!), but before about AD 200 Christmas fell in Capricorn, and it was over the four weeks preceding December 25 that the Sun was in Sagittarius – hence the name 'Advent grains'.

We started late-sowing trials in 1963, sowing rye daily throughout December and January. If the ground was frozen, we laid the seed on top and covered it with chaff; if there was snow, we sowed into the snow. The results varied; some of the plants sown in the first half of December (while the Sun was in Scorpio) suffered fungal attacks in spring when the Sun entered Pisces. Sowings made when the Sun moved into Capricorn (around January 20) tillered well but did not come into ear; they need a further growing period in order to form seeds.

Incidentally, when I showed my father – by then almost ninety years old – the results of these trials, he told me that his father had been a 'sower'. He always

17

Walter Thun, The Twelvefold Human Being, *Pen and ink*

sowed grain, not only on his own farm but also on neighbouring farms, during Advent, a task he later passed on to his son. After the First World War the idea was frowned upon and few farmers asked my father to sow Advent grain. The new fertilising techniques made such considerations irrelevant.

We continued our sowing trials for three years, always growing a crop from the saved seeds the year after. Other biodynamic researchers like Erika Windeck, Martin Schmidt and Erhard Breda were conducting similar trials. They were apparently searching for 'good Full Moon sowings'. Their experience led them to conclude however that the midwinter New Moon provided the best conditions for the next year's seeds. They were of course only looking at the phase of the Moon, not its position in the zodiac. We were able to share our findings with them, namely that first-class seed was produced when both Sun and Moon were in Sagittarius (although we also had good results when the Sun was in Sagittarius and the Moon in Aries or Leo).

Martin Schmidt, with whom we communicated regularly, had felt Virgo to be the ideal constellation for grain; it is after all associated with the goddess Demeter who holds an ear of grain – the star of Spica – in her hand. Afterwards Schmidt started to explore other zodiac influences as well.

Erika Windeck later undertook sowing trials with wall barley and rye brome, and we often compared results. We also discussed these issues with Theodor Schwenk, Agnes Fyfe, Suso Vetter, Rudolf Hauschka, Udo Renzenbrinck, Walter Bühler, Wilhelm Pelikan and others.

Returning to the question regarding the influence of phases of the Moon, it's interesting to study rice, a plant intimately connected with water and grown in paddy fields. A strong connection to lunar phases in this case would make sense. Lili Kolisko did a lot of research into Moon phases and concluded that two extreme situations were possible. If the Full Moon influence is too powerful the crop has a tendency to rot; if the New Moon influence is too strong there is a tendency towards lignification, for becoming woody. Our experience over decades has born this out, although it's worth remembering that Rudolf Steiner stated clearly that plants should grow in enlivened soil, not enlivened water.

If our soils are truly enlivened in this way the Moon is able to not only work on earthly substances through its phases, but is able like the planets to act as mediator for the more distant forces of the twelve constellations of the zodiac. If we can harness that power through our care of the soil, we should be able to produce nutritional food for human beings which will in turn nourish the body in its twelvefold nature. In this way, eating can itself become an act of creation.

Best Practice for Harvesting and Storing Seed

Maria Thun

If we do manage to care for our soil in a way that allows the cosmic forces to work through it, we are faced with the issue of seed quality. The ability to genetically modify plants has raised many questions about seed plants and biodiversity in recent years. A number of different seed initiatives have emerged in recent years dedicated to the preservation of old varieties or the selection of new ones. Results have been mixed due to differing approaches – some remain committed to observing lunar phases while others focus on different kinds of manuring.

Since plant breeding is a complex and difficult undertaking, we would like to focus on the producton of already existing good quality seed varieties. Root vegetables make an excellent case study because their reproduction is relatively simple. The easiest to produce seed from are radishes. They work well because, if sown early enough, they can produce seed by late autumn. The plants need to be thoroughly dried before threshing.

A basic rule for good seed production is to allow the plant to grow until it has begun to develop the desired fruiting body. In the case of a root plant this is of course the root. (A leaf plant like salad needs to develop a head; tomatoes or

A clamp of beets showing leaf stems

cucumbers need to show the beginnings of these fruits, similarly with flowers, and so on.) Once that has happened, the approach is then changed and all subsequent cultivations and spraying should be carried out at Fruit times in order to support good seed development. Harvesting should be done at Root times.

If seed is to be collected the following year from carrots, beetroot or turnips, they will need to be stored over winter. A cool cellar is suitable for this purpose; if a large quantity of roots is to be stored, use a clamp outdoors. The clamp should be covered with straw. Rye straw is ideal because it doesn't degrade easily, even when it's on the ground. Finally, cover the straw with 20 cm (8 in) of soil; in areas with high snowfall or cold frosts, spruce or fir branches can also be laid on top to provide an air cushion for extra insulation. Smaller quantities of roots (as well as apples and pears) can be stored over winter in the barn or a loft with a covering of straw.

A good seed head needs to develop from the roots when they are planted in spring. Care should therefore be taken to leave 3–4 cm (1–1½ in) of the old leaves attached when cutting off the tops. Leaves may also be twisted off but again take care not to damage the crown. If this happens (as is frequently the case with mechanised harvesting) several small flower shoots are likely to develop around the edges of the cut with no central shoot being formed. This affects the quality of the seed. Secondary roots should not be removed. Seeding shoots that have started to form in the first year should not be used. Their seeds are not

Twisted off leaves; the leaf core of a carrot is clearly visible

Beetroot with twisted off leaves and visible leaf core

Carrot inflorescence, unusable for seeds as it is from the first year

Carrot flower

Carrot inflorescence

Carrot seeds

Beetroot

Beetroot inflorescence

Beetroot infructescence

much use and will result in poor quality fruits. Intensively bred industrial juicing carrots are also unsuitable for seed and have often lost the typical carrot flavour.

The roots which would have been harvested in the autumn at Root times should be replanted in spring at Fruit times to encourage them to invest their strength in seed production. For the best quality, choose times when the Moon is in Leo. The fully grown seed heads should be harvested at Root times once the seed has ripened. The seeds will then grow good roots the following year.

Fodder beet

Fodder beet infructescence

Background to the Calendar

The zodiac

The **zodiac** is a group of twelve constellations of stars which the Sun, Moon and all the planets pass on their circuits. The Sun's annual path always takes exactly the same line, called **ecliptic.** The Moon's and planets' paths vary slightly, sometimes above and sometimes below the ecliptic. The point at which their paths cross the ecliptic is called a **node** (☊ and ☋).

The angles between the Sun, Moon and planets are called **aspects.** In this calendar the most important is the 120° angle, or **trine.**

In the illustration below the outer circle shows the varying sizes of the visible **constellations** of the **zodiac.** The dates on this outer circle are the days on which the Sun enters the constellation (this can change by one day because of leap years). The inner circle shows the divisions into equal sections of 30° corresponding to the **signs** used in astrology.

It is the *constellations,* not the signs, on which our observations are based, and which are used throughout this calendar.

The twelve constellations are grouped into four different types, each having three constellations at an angle of about 120°, or trine. About every nine days the Moon passes from one type, for instance Root, through the other types (Flower, Leaf and Fruit) and back to Root again.

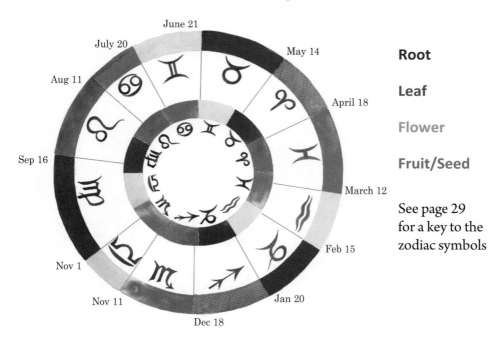

Root

Leaf

Flower

Fruit/Seed

See page 29 for a key to the zodiac symbols

24

If a New Moon is at a node there is a solar eclipse, as the Moon is directly in front of the Sun, while a Full Moon at a node causes a lunar eclipse where the Earth's shadow falls on the Moon. If the Sun or Moon pass exactly in front of a planet, there is an occultation (•). If Mercury or Venus pass exactly in front of the Sun, this is a transit (other planets cannot pass in front of the Sun).

What are oppositions, trines and conjunctions?
Oppositions ☍

A **geocentric** (Earth-centred) **opposition** occurs when for the observer on the Earth there are two planets opposite one another – 180° apart – in the heavens. They look at one another from opposite sides of the sky and their light interpenetrates. Their rays fall on to the Earth and stimulate in a beneficial way the seeds that are being sown in that moment. In our trials we have found that seeds sown at times of opposition resulted in a higher yield of top quality crops.

With a **heliocentric** (Sun-centred) **opposition** an observer would need to place themselves on the Sun. This is of course physically impossible but we can understand it through our thinking. The Sun is in the centre and the two planets placed 180° apart also gaze at each other but this time across the Sun. Their rays are also felt by the Earth and stimulate better plant growth. However, heliocentric oppositions are not shown or taken into account in the calendar.

At times of opposition two zodiac constellations are also playing their part. If one planet is standing in a Warmth constellation, the second one will usually be in a Light constellation or vice versa. If one planet is in a Water constellation, the other will usually be in an Earth one. (As the constellations are not equally sized, the point opposite may not always be in the opposite constellation.)

Trines △ or ▲

The twelve constellations are grouped into four different types, each having three constellations at an angle of about 120°, or trine. About every nine days the Moon passes a similar region of forces.

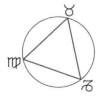
Earth-Root

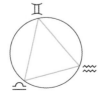
Light-Flower

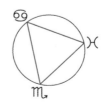
Water-Leaf

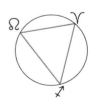
Warmth-Fruit/Seed

Trines occur when planets are 120° from one another. The two planets are then usually both standing in the same elemental configuration – Aries and Leo for example are both Warmth constellations. A Warmth trine means that the effects of these constellations will enhance fruit and seed formation in the plants sown at this time. If two planets are in trine position in Water, watery influences will be enhanced, which usually brings high rainfall. Plants sown on these days will yield more leaf than those on other days. Trine effects can change the way plants grow.

Conjunctions ☌

Conjunctions and multiple conjunctions occur when two or more planets stand behind one another in space. It is then usually only the planet closest to the Earth that has any influence on plant growth. If this influence is stronger than that of the sidereal Moon, cosmic disturbances can occur that irritate the plant and cause checks in growth. This negative effect is increased further when the Moon or another planet stands directly in front of another – an occultation (•) or eclipse in the case of Sun and Moon. Sowing at these times will affect subsequent growth detrimentally and harm a plant's regenerative power.

The effects of the Moon

In its 27-day orbit round the Earth the Moon passes through the constellations of the zodiac and transmits forces to the Earth which affect the four elements: Earth, Light (Air), Water and Warmth (Fire). They in turn affect the four parts of the plant: the roots, the flower, the leaves and the fruit or seeds. The health and growth of a plant can therefore be stimulated by sowing, cultivating and harvesting it in tune with the cycles of the Moon.

These cosmic forces can also be harnessed in beekeeping. By opening and closing the bee 'skep' or box in rhythm with the Moon, the bees' activity is directly affected.

The table opposite summarises the effects of the movement of the Moon through the twelve constellations on plants, bees and the weather.

The amount of time the Moon spends in any constellation varies between two and four days. However, this basic framework can be disrupted by planetary oppositions which override the normal tendencies; equally, it may be that trine positions (see above) activate a different elemental force to the ones the Moon is transmitting. Times when the Moon's path or a planet's path intersects with the ecliptic (ascending ☊ or descending ☋ node; see page 24) are subject to mainly negative effects. These are intensified if there is an eclipse or occultation, in which case the nearer planet interrupts the influence of the distant one. Such days are unsuitable for sowing or harvesting.

Constellation	Sign	Element	Plant	Bees	Weather
Pisces, Fishes	♓ W	Water	Leaf	Making honey	Damp
Aries, Ram	♈ H	Warmth	Fruit	Gathering nectar	Warm/hot
Taurus, Bull	♉ E	Earth	Root	Building comb	Cool/cold
Gemini, Twins	♊ L	Light	Flower	Gathering pollen	Airy/bright
Cancer, Crab	♋ W	Water	Leaf	Making honey	Damp
Leo, Lion	♌ H	Warmth	Fruit	Gathering nectar	Warm/hot
Virgo, Virgin	♍ E	Earth	Root	Building comb	Cool/cold
Libra, Scales	♎ L	Light	Flower	Gathering pollen	Airy/bright
Scorpio, Scorpion	♏ W	Water	Leaf	Making honey	Damp
Sagittarius, Archer	♐ H	Warmth	Fruit	Gathering nectar	Warm/hot
Capricorn, Goat	♑ E	Earth	Root	Building comb	Cool/cold
Aquarius, Waterman	♒ L	Light	Flower	Gathering pollen	Airy/bright

Groupings of plants for sowing and harvesting

When we grow plants, different parts are cultivated for food. We can divide them into four groups.

Root crops at Root times

Radishes, swedes, sugar beet, beetroot, celeriac, carrot, scorzonera, etc., fall into the category of root plants. Potatoes and onions are included in this group too. Root times produce good yields and top storage quality for these crops.

Leaf plants at Leaf times

The cabbage family, lettuce, spinach, lambs lettuce, endive, parsley, leafy herbs and fodder plants are categorised as leaf plants. Leaf times are suitable for sowing and tending these plants but not for harvesting and storage. For this (as well as harvesting of cabbage for sauerkraut) Fruit and Flower times are recommended.

Flower plants at Flower times

These times are favourable for sowing and tending all kinds of flower plants but also for cultivating and spraying 501 (a biodynamic preparation) on oil-bearing plants such as linseed, rape, sunflower, etc. Cut flowers have the strongest scent and remain fresh for longer if cut at Flower times, and the mother plant will provide many new side shoots. If flowers for drying are harvested at Flower times they retain the most vivid colours. If cut at other times they soon lose their colour. Oil-bearing plants are best harvested at Flower times.

Fruit Plants at Fruit times

Plants that are cultivated for their fruit or seed belong to this category, including beans, peas, lentils, soya, maize, tomatoes, cucumber, pumpkin, zucchini, but also cereals for summer and winter crops. Sowing oil-bearing plants at Fruit times provides the best yields of seeds. The best time for extraction of oil later on is at Flower times. Leo times are particularly suitable to grow good seed. Fruit plants are best harvested at Fruit times. They store well and their seeds provide good plants for next year. When storing fruit, also remember to choose the time of the ascending Moon.

There is always uncertainty as to which category some plants belong (see list on p. 57). Onions and beetroot provide a similar yield when sown at Root and Leaf times, but the keeping quality is best from Root times. Kohlrabi and cauliflowers belong to Leaf times, as does Florence fennel. Broccoli is more beautiful and firmer when sown at Flower times.

Explanations of the Calendar Pages

Next to the date is the constellation (and time of entry) in which the Moon is positioned. This is the astronomical constellation, not the astrological sign (see p. 24). The next column shows solar and lunar events.

A further column shows which element is dominant on that day (this is useful for beekeepers). Note H is used for warmth (heat). Sometimes there is a change during the day; in this case, both elements are mentioned. Warmth effects on thundery days are implied but are not mentioned in this column, but may have a ⼁ symbol in the far right 'Weather' column.

The next column shows in colour the part of the plant which will be enhanced by sowing or cultivation on that day. Numbers indicate times of day. On the extreme right, special events in nature are noted as well as anticipated weather changes which disturb or break up the overall weather pattern.

When parts of the plant are indicated that do not correspond to the Moon's position in the zodiac (often it is more than one part on the same day), it is not a misprint, but takes account of other cosmic aspects which overrule the Moon-zodiac pattern and have an effect on a different part of the plant.

Unfavourable times are marked thus ▇. These are caused by eclipses, nodal points of the Moon or the planets or other aspects with a negative influence; they are not elaborated upon in the calendar. If one has to sow at unfavourable times for practical reasons, one can choose favourable days for hoeing, which will improve the plant.

The position of the planets in the zodiac is shown in the box below, with the date of entry into a new constellation. R indicates the planet is moving retrograde (with the date when retrograde begins), D indicates the date when it moves in direct motion again.

On the opposite calendar page astronomical aspects are indicated. Those visible to the naked eye are shown in **bold** type. Visible conjunctions (particularly Mercury's) are not always visible from all parts of the Earth.

Astronomical symbols

Constellations		Planets		Aspects			
♓	Pisces	☉	Sun	☊	Ascending node	**St**	Storms likely
♈	Aries	☾, ☽	Moon	☋	Descending node	♄	Thunder likely
♉	Taurus	☿	Mercury	⌢	Highest Moon	**Eq**	Earthquakes
♊	Gemini	♀	Venus	⌣	Lowest Moon	**Tr**	Traffic dangers
♋	Cancer	♂	Mars	**Pg**	Perigee	**Vo**	Volcanic activity
♌	Leo	♃	Jupiter	**Ag**	Apogee		Northern Transplanting Time
♍	Virgo	♄	Saturn	☍	Opposition		
♎	Libra	♅	Uranus	☌	Conjunction		
♏	Scorpio	♆	Neptune	☄	Eclipse/occultation		Southern Transplanting Time
♐	Sagittarius	♇	Pluto	☄	Lunar eclipse		
♑	Capricorn	○	Full Moon	△	Trine (or ▲)		
♒	Aquarius	●	New Moon	E Earth	L Light/Air	W Water	H Warmth/Heat

Transplanting times

From midwinter through to midsummer the Sun rises earlier and sets later each day while its path across the sky ascends higher and higher. From midsummer until midwinter this is reversed, the days get shorter and the midday Sun shines from an ever lower point in the sky. This annual ascending and descending of the Sun creates our seasons. As it ascends and descends during the course of the year the Sun is slowly moving (from an Earth-centred point of view) through each of the twelve constellations of the zodiac in turn. On average it shines for one month from each constellation.

In the northern hemisphere the winter solstice occurs when the Sun is in the constellation of Sagittarius and the summer solstice when it is in Gemini. At any point from Sagittarius to Gemini the Sun is ascending, while from Gemini to Sagittarius it is descending. In the southern hemisphere this is reversed.

The Moon (and all the planets) follow approximately the same path as the Sun

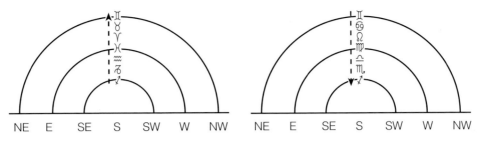

Northern hemisphere ascending Moon (left) and descending Moon (right): Transplanting Time

around the zodiac but instead of a year, the Moon takes only about 27½ days to complete one cycle, shining from each constellation in turn for a period of two to three days. This means that the Moon will ascend for about fourteen days and then descend.

It is important to distinguish the journey of the Moon through the zodiac (siderial rhythm) from the waxing and waning (synodic) cycle: in any given constellation there may be a waxing, waning, full, quarter, sickle or gibbous Moon. As it moves through the zodiac the Moon, like the Sun, is ascending (in the northern hemisphere) when it is in the constellations from Sagittarius to Gemini and descending from Gemini to Sagittarius. In the southern hemisphere it is ascending from Gemini to Sagittarius and descending from Sagittarius to Gemini.

When the Moon is ascending, plant sap rises more strongly. The upper part of the plant fills with sap and vitality. This is a good time for cutting scions (for grafting). Fruit harvested during this period remains fresh for longer when stored.

When the Moon is descending, plants take root readily and connect well with their new location. This period is referred to as the **Transplanting Time.** Moving plants from one location to another is called *transplanting*. This is the case when young plants are moved from the seed bed into their final growing position but also when the gardener wishes to strengthen the root development of young fruit trees, shrubs or pot plants by frequently re-potting them. Sap movement is slower during the descending Moon. This is why it is a good time for trimming hedges, pruning trees and felling timber as well as applying compost to meadows, pastures and orchards.

Note that sowing is the moment when a seed is put into the soil; either the ascending or descending period can be used. It then needs time to germinate and grow. This is different from *transplanting*, which is best done during the descending Moon. These times given in the calendar. Northern Transplanting Times refer to the northern hemisphere, and **Southern Transplanting Times** refer to the southern hemisphere. All other constellations and planetary aspects are equally valid in both hemispheres.

Local times

Times given are *Eastern Standard Time* (EST), or from March 12 to Nov 4 *Eastern Daylight Saving Time* (EDT), with ^{a} or $_p$ after the time for am and pm.

Noon is 12_p and midnight is 12^a; the context shows whether midnight at the beginning of the day or at the end is meant; where ambiguous (as for planetary aspects) the time has been adjusted by an hour for clarity.

For different time zones adjust as follows:

Newfoundland Standard Time: add $1\frac{1}{2}^h$
Atlantic Standard Time: add 1^h
Eastern Standard Time: do not adjust
Central Standard Time: subtract 1^h
 For Saskatchewan subtract 1^h, but subtract 2^h from March 12 to Nov 4 (no DST)
Mountain Standard Time: subtract 2^h
 For Arizona subtract 2^h, but subtract 3^h from March 12 to Nov 4 (no DST)
Pacific Standard Time: subtract 3^h
Alaska Standard Time: subtract 4^h
Hawaii Standard Time: subtract 5^h, but subtract 6^h from March 12 to Nov 4 (no DST)

For Central & South America adjust as follows:

Argentina: add 2^h, but add 1^h from March 12 to Nov 4 (no DST)
Brazil (Eastern): Jan 1 to Feb 18 add 3^h; Feb 19 to March 11 add 2^h; March 12 to Oct 14 add 1^h; Oct 15 to Nov 4 add 2^h; from Nov 5 add 3^h.
Chile: Jan 1 to March 11 add 2^h; March 12 to May13 add 1^h; May 14 to Aug 12 do not adjust; Aug 13 to Nov 4 add 1^h; from Nov 5 add 2^h.
Columbia, Peru: do not adjust, but subtract 1^h from March 12 to Nov 4 (no DST).
Mexico (mostly CST): subtract 1^h, but from March 12 to April 3, and from Oct 29 to Nov 4, subtract 2^h

For other countries use *The Maria Thun Biodynamic Calendar* from Floris Books which carries all times in GMT, making it easier to convert to another country's local and daylight saving time.

January 2017

All times in EST

Date	Const. of Moon		Solar & lunar aspects	Trines	Moon El'ment	Parts of the plant enhanced by Moon or planets (12 1 2 3 4 5 6 7 8 9 10 11 12 1 2 3 4 5 6 7 8 9 10 11 12)	Weather
1 Sun	♒	11_p	☉-♐		E/L	Root to 10_p | 11	Tr
2 Mon	♒		☋ 1_p	☽♆	L	Flower to 9^a | (unlabelled bar) | Fl 5 - 8_p | (unlabelled bar)	
3 Tue	♓	9_p		☽♂	L/W	(unlabelled bar) | Flower from 3^a to 8_p | Lf 9_p	
4 Wed	♓				W	Leaf	Tr
5 Thu	♓		☽ 3_p		W	Leaf	St Vo
6 Fri	♈	3_p			W/H	Leaf to 2_p | Fruit from 3_p	
7 Sat	♈				H	Fruit	♄ Tr
8 Sun	♉	7^a			H/E	Fruit to 6^a | Root from 7^a	♄ St
9 Mon	♉				E	Root to 1_p | (unlabelled bar)	St
10 Tue	♊	6_p	**Pg** 1^a		E/L	(unlabelled bar) | Rt 1_p-5_p | Flower 6_p	♄ St
11 Wed	♊		⌒ 5^a		L	Flower	
12 Thu	♋	4_p	○ 7^a		L/W	Flower to 3_p | Leaf from 4_p	
13 Fri	♋				W	Leaf	
14 Sat	♌	3^a			W/H	-2^a | Fruit from 3^a	
15 Sun	♌		☊ 6^a		H	-2^a | (unlabelled bar) | Fruit from 10^a	St Eq
16 Mon	♍	7_p			H/E	Fruit to 5^a | (unlabelled bar)	
17 Tue	♍			♀☊	E	(unlabelled bar)	♄ St
18 Wed	♍		☉-♑		E	Root from 1^a	Tr
19 Thu	♍		☾ 5_p		E	Root	St
20 Fri	♎	1_p			E/L	Root to 12_p | Flower from 1_p	
21 Sat	♎		**Ag** 7_p		L	Flower to 12^a	
22 Sun	♏	1^a			W	Leaf from 1^a	
23 Mon	♏				W	Leaf	
24 Tue	♐	3_p			W/H	Leaf to 2_p | Fruit from 3_p	St Eq
25 Wed	♐		◡ 7^a		H	Fruit	
26 Thu	♐				H	Fruit to 12^a	
27 Fri	♑	1^a	● 7_p		E	(unlabelled bar)	St
28 Sat	♑			☿☋	E	(unlabelled bar) | Root from 2_p	♄ Vo Tr
29 Sun	♒	5^a	☋ 5_p		E/L	Root to 4^a | Flower 5^a to 1_p | (unlabelled bar) | 10_p	St
30 Mon	♒			☽♆	L	Fl to 4^a | (unlabelled bar) | Flower from 9^a	
31 Tue	♓	2^a			L/W	1^a | Leaf from 2^a	Tr

Northern Transplanting Time (11 Wed – 24 Tue)

12 1 2 3 4 5 6 7 8 9 10 11 12 1 2 3 4 5 6 7 8 9 10 11 12

Mercury ☿	Venus ♀	Mars ♂	Jupiter ♃	Saturn ♄	Uranus ♅	Neptune ♆	Pluto ♇
♐	♒	♒	♍	♏	♓	♒	♐
(R 8 D)	24 ♓	17 ♓					

NB: All zodiac symbols refer to astronomical constellations, not astrological signs (see p. 24)

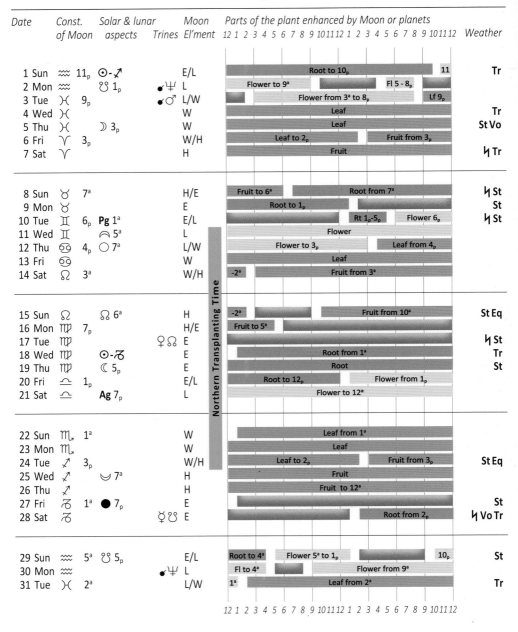

Planetary aspects
(**Bold** = visible to naked eye)

January 2017

1	♂☌♆ 2ᵃ
2	**☽☌♀ 3ᵃ** ☽●♆ 11ₚ
3	☽●♂ 2ᵃ
4	
5	☽☌⊕ 11ₚ
6	☽☌♃ 1ᵃ
7	☉☌♇ 2ᵃ
8	
9	
10	☽☍♄ 6ᵃ ☽☍☿ 5ₚ
11	☽☌♇ 10ₚ
12	♀☌♆ 5ₚ
13	
14	
15	☾☍♆ 5ₚ ☾☍♀ 11ₚ
16	☾☍♂ 2ₚ
17	♀☍☊ 1ₚ
18	☾☍⊕ 11ₚ
19	☾☌♃ 2ᵃ
20	
21	
22	
23	
24	☾☌♄ 6ᵃ
25	☾☌☿ 7ₚ
26	☾☌♇ 4ᵃ
27	
28	☿☍☊ 1ᵃ
29	☿☌♇ 3ₚ
30	☽●♆ 6ᵃ
31	☽☌♀ 1ₚ ☽☌♂ 10ₚ

At the beginning of the month Mercury, Venus and Mars, supported by Neptune and Pluto, will bring warmth and light. From Jan 17 Water and cool Earth influences will try to work against Mercury bringing cool precipitation.

Northern Transplanting Time
Jan 11 7ᵃ to Jan 25 5ᵃ
Southern Transplanting Time
Dec 29 to Jan 11 2ᵃ and
Jan 25 9ᵃ to Feb 7

The transplanting time is a good time for **pruning fruit trees, vines and hedges.** Fruit and Flower times are preferred for this work. Avoid unfavourable times.

When **milk processing** it is best to avoid unfavourable times. This applies to both butter and cheese making. Milk which has been produced at Warmth/Fruit times yields the highest butterfat content. This is also the case on days with a tendency for thunderstorms. Times of moon perigee (**Pg**) are almost always unfavourable for milk processing and even yoghurt will not turn out well. Starter cultures from such days decay rapidly and it is advisable to produce double the amount the day before. Milk loves Light and Warmth times best of all. Water times are unsuitable.

Southern hemisphere harvest time for seeds
Fruit seeds: Jan 14 3ᵃ to Jan 16 5ᵃ and at other Fruit times, always avoiding unfavourable times.
Flower seeds: Flower times.
Leaf seeds: Leaf times.
Root seeds: Root times.

Control slugs from Jan 12 4ₚ to Jan 14 2ᵃ.

Planet (naked eye) visibility
Evening: Venus, Mars
All night: Jupiter
Morning: Mercury (from 4th) , Saturn

▬▬▬▬ *Unfavourable time*   33

February 2017

Date	Const. of Moon	Solar & lunar aspects	Trines	Moon El'ment	Parts of the plant enhanced by Moon or planets 12 1 2 3 4 5 6 7 8 9 10 11 12 1 2 3 4 5 6 7 8 9 10 11 12	Weather
1 Wed ♓		⊙-♐		W	Leaf	♄ St
2 Thu ♈	9ₚ			W/H	Leaf to 8ₚ Fr 9ₚ	St
3 Fri ♈		☽ 11ₚ		H	Fruit	St Vo Tr
4 Sat ♉	1ₚ			H/E	Fruit to 12ₚ Root from 1ₚ	St Eq
5 Sun ♉				E	Root to 9ₚ	
6 Mon ♉		Pg 9ᵃ		E	Rt 9ₚ	
7 Tue ♊	2ᵃ ♑ 2ₚ			E/L	1ᵃ Flower from 2ᵃ	
8 Wed ♊				L	Flower to 12ᵃ	St
9 Thu ♋	1ᵃ			W	Leaf from 1ᵃ	
10 Fri ♌	1ₚ ⊙🌕○ 8ₚ			W/H	Leaf to 12ₚ 1ₚ - 4ₚ 10ₚ	St Eq
11 Sat ♌	♌ 3ₚ		▲	H	Root to 11ᵃ Fruit from 6ₚ	
12 Sun ♌				H	Fruit	St Vo
13 Mon ♍	4ᵃ			H/E	Fr - 3ᵃ Root from 4ᵃ	♄ Eq
14 Tue ♍				E	Root	St
15 Wed ♍		⊙-♒		E	Root	
16 Thu ♎	9ₚ			E/L	Root to 8ₚ Fl 9ₚ	
17 Fri ♎				L	Flower	
18 Sat ♏	10ᵃ ☾ 3ₚ Ag 4ₚ		L/W	Flower to 8ₚ Lf 9ₚ	
19 Sun ♏				W	Leaf	
20 Mon ♏				W	Leaf to 11ₚ	St Eq
21 Tue ♐	12ᵃ ☌ 4ₚ		▲	H	12-2ᵃ Root from 3ᵃ to 5ₚ Fruit from 6ₚ	St
22 Wed ♐				H	Fruit	St Vo
23 Thu ♑	10ᵃ			H/E	Fruit to 9ᵃ Root from 10ᵃ	
24 Fri ♑				E	Root	St
25 Sat ♒	2ₚ			E/L	Root to 1ₚ Flower 2ₚ to 10ₚ	♄ St
26 Sun ♒	☋ 1ᵃ ⊙🌑● 10ᵃ		L		♂♅
27 Mon ♓	10ᵃ	♂♌		L/W	Tr
28 Tue ♓				W	Leaf from 1ᵃ	St Vo

12 1 2 3 4 5 6 7 8 9 10 11 12 1 2 3 4 5 6 7 8 9 10 11 12

Northern Transplanting Time

Mercury ☿	Venus ♀	Mars ♂	Jupiter ♃	Saturn ♄	Uranus ♅	Neptune ♆	Pluto ♇
♐ 6 ♑	♓	♓	♍	♏	♓	♒	♐
23 ♒			(6 R)				

NB: All zodiac symbols refer to astronomical constellations, not astrological signs (see p. 24)

Planetary aspects
(Bold = *visible to naked eye)*

February 2017

1	
2	☽♂♁ 6ᵃ ☽♂♃ 9ᵃ
3	
4	
5	
6	☽♂♄ 6ₚ
7	
8	☽♂♇ 9ᵃ
9	☽♂☿ 10ᵃ
10	
11	☉△♃ 10ᵃ
12	☾♂♆ 5ᵃ
13	
14	☾♂♀ 6ᵃ ☾♂♂ 5ₚ
15	☾♂♁ 9ᵃ ☾♂♃ 12ₚ
16	
17	
18	
19	
20	☾♂♄ 7ₚ
21	☿△♃ 1ₚ
22	☾♂♇ 3ₚ
23	
24	
25	☾♂☿ 8ₚ
26	☽●♆ 4ₚ ♂♂♁ 7ₚ
27	♂♊ 1ᵃ ♂♂♃ 9ᵃ
28	☽♂♀ 10ₚ

In the first week Mercury is still in the warm constellation of Sagittarius, but then moves to cool Capricorn. Supported by Jupiter in Virgo, this could bring some cold against the predominant Water influences.

Northern Transplanting Time
Feb 7 4ₚ to Feb 21 2ₚ
Southern Transplanting Time
Jan 25 to Feb 7 12ₚ and
Feb 21 6ₚ to March 6

Vines, fruit trees and shrubs can be pruned during Transplanting Time selecting Flower and Fruit times in preference. Avoid unfavourable times.

Best times for taking **willow cuttings for hedges and fences:** At Flower times outside Transplanting Time. In warm areas at Flower times during Transplanting Time to avoid too strong a sap current.

Southern hemisphere harvest time for seeds
Fruit seeds: Feb 10 1ₚ to 4ₚ, and Feb 11 6ₚ to Feb 13 3ᵃ and at other Fruit times.
Flower seeds: Feb 7 2ᵃ to Feb 8 11ₚ and at other Flower times.

Control slugs from Feb 9 1ᵃ to Feb 10 12ₚ.

Planet (naked eye) visibility
Evening: Venus, Mars
All night: Jupiter
Morning: Mercury (to 2nd) , Saturn

March 2017

All times in EST / EDT

Date	Const. of Moon	Solar & lunar aspects	Trines	Moon El'ment	Parts of the plant enhanced by Moon or planets	Weather
1 Wed	♓	☉-♒		W	Leaf	♄ St
2 Thu	♈ 3^a			W/H	-2^a; Fruit from 3^a to 3_p	St Vo
3 Fri	♉ 7_p	**Pg** 2^a		H/E	3_p-6_p; Root 7_p	
4 Sat	♉			E	Root	Tr
5 Sun	♉	☽ 7^a	▲	E	Rt -3^a; Leaf from 4^a to 8_p; Rt 9_p	St Eq
6 Mon	♊ 8^a	⌒ 8_p		E/L	Root to 7^a; Flower from 8^a	♄ St
7 Tue	♊			L	Flower	
8 Wed	♋ 8^a			L/W	Flower to 7^a; Leaf from 8^a	
9 Thu	♌ 9_p			W/H	Leaf to 8_p; Fr 9_p	
10 Fri	♌	☊ 11_p		H	Fruit to 8_p	
11 Sat	♌	☉-♓		H	Fruit from 4^a	St
DST begins. All times below in ***EDT***						
12 Sun	♍ 2_p	○ 11^a		H/E	Fruit to 1_p; Root from 2_p	St
13 Mon	♍			E	Root	
14 Tue	♍			E	Root	Vo Tr
15 Wed	♍			E	Root	
16 Thu	♎ 7^a			E/L	Root to 6^a; Flower from 7^a	
17 Fri	♏ 7_p			L/W	Flower to 6_p	St
18 Sat	♏	**Ag** 1_p	☿☊	W		
19 Sun	♏			W	Leaf from 8^a	
20 Mon	♐ 9^a	☾ 12_p		W/H	Leaf to 8^a; Fruit from 9^a	
21 Tue	♐	◡ 1^a		H	Fruit	Vo Tr
22 Wed	♑ 8_p			H/E	Fruit to 7_p; Root 8_p	
23 Thu	♑			E	Root	St Eq
24 Fri	♑			E	Root to 11_p	St Vo
25 Sat	♒ 12^a	☋ 12_p		L	Flower 12^a to 8^a; Flower 3_p	St
26 Sun	♓ 8_p		☌♆	L/W	-2^a; Flower from 7^a to 7_p; Leaf 8_p	
27 Mon	♓	● 11_p		W	Leaf	St
28 Tue	♓			W	Leaf	
29 Wed	♈ 12_p		▲	W/H	Leaf to 5_p; Fr 6-9_p	St Eq
30 Thu	♈	**Pg** 8^a		H	10_p	St Vo
31 Fri	♉ 3^a			H/E	-2^a; Root from 3^a	Vo Tr

Northern Transplanting Time (6–20 March)

Mercury ☿	Venus ♀	Mars ♂	Jupiter ♃	Saturn ♄	Uranus ⛢	Neptune ♆	Pluto ♇
♒ 9 ♓	♓	♓	♍	♏	♓	♒	♐
31 ♈	(4 R)	9 ♈	(R)				

NB: All zodiac symbols refer to astronomical constellations, not astrological signs *(see p. 24)*

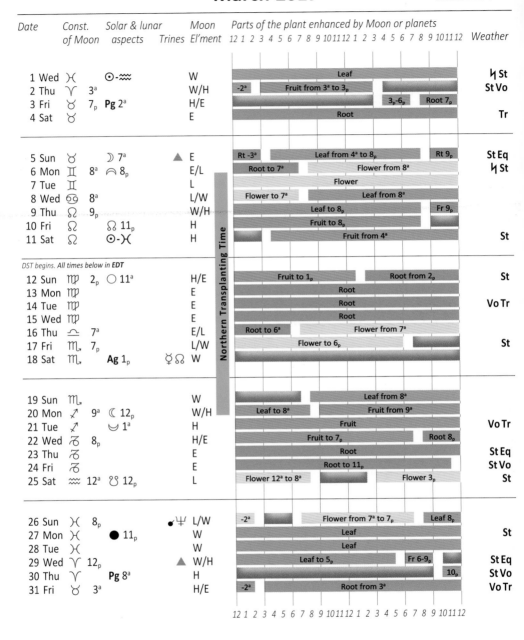

Planetary aspects
(**Bold** = *visible to naked eye*)

March 2017

1 ☽♂⚴ 1p ☽♂♃ 2p **☽♂♂ 5p** ☉♂♆ 10p
2 ♃♂⚴ 9p
3
4 ☿♂♆ 6a

5 ♂△♄ 4p
6 ☽♂♄ 3a ☉♂☿ 8p
7 ☽♂♇ 4p
8
9
10
11 ☽♂♆ 3p

12 ☾♂♂☿ 11p
13
14 ☾♂♀ 1a ☾♂♃ 6p ☾♂⚴ 9p
15 ☾♂♂ 8p
16
17
18 ☿♂♀ 8a ☿☊ 6p

19
20 ☾♂♄ 7a
21
22 ☾♂♇ 2a
23
24 ☿♂♃ 9a
25 ☉♂♀ 6a

26 ☾♂♆ 4a ☿♂⚴ 11a
27 ☾♂♀ 4p
28 ☽♂♃ 7p
29 ☽♂⚴ 1a ☽♂☿ 8a ☿△♄ 2p
30 ☽♂♂ 12p
31

Will Mercury bring some warmth and light in the first week? From March 9 there will be much precipitation. Will Jupiter retrograde in Virgo bring some wintry cold?

Northern Transplanting Time
March 6 10p to March 20 11p
Southern Transplanting Time
Feb 21 to March 6 6p and
March 21 4a to April 2

Willow cuttings for **pollen production** are best cut from March 16 7a to March 17 6p; and for **honey flow** from March 9 9p to March 12 1p. Avoid unfavourable times.

The cuttings taken in February are best stuck in the ground during transplanting time; to improve pollen production do this at Flower times, and to increase honey flow do this at Fruit times.

Control slugs from March 8 8a to March 9 8p.

Cuttings for grafting: Cut outside Transplanting Time during ascending Moon – always choosing times (Fruit, Leaf, etc.) according to the part of plant to be enhanced.

Southern hemisphere harvest time for seeds
Fruit seeds: March 9 9p to March 12 1p and at other Fruit times.
Flower seeds: March 6 8a to March 8 7a and at other Flower times.
Leaf seeds: March 8 8a to March 9 8p and at other Leaf times.
Root seeds: Root times.
Always avoid unfavourable times.

Biodynamic preparations: Pick dandelion in March or April in the mornings during Flower times. The flowers should not be quite open in the centre. Dry them on paper in the shade, not in bright sunlight.

Planet (naked eye) visibility
Evening: Mercury (from 18th), Venus (to 23rd), Mars
All night: Jupiter
Morning: Venus (from 21st), Saturn

Unfavourable time

Date	Const. of Moon	Solar & lunar aspects	Trines	Moon El'ment	Parts of the plant enhanced by Moon or planets	Weather
1 Sat	♉	☉-♓		E	Root	St
2 Sun	♊ 12p			E/L	Root to 1p / Flower from 2p	
3 Mon	♊	⌒ 2a ☽ 3p		L	Flower	St Eq
4 Tue	♋ 12p			L/W	Flower to 1p / Leaf from 2p	
5 Wed	♋		▲	W	Leaf to 12p / Fruit from 1p	
6 Thu	♌ 4a			W/H	Fruit	St
7 Fri	♌	☍ 5a		H	1a / Fruit from 10a	
8 Sat	♍ 9p			H/E	Fruit to 8p / Rt 9p	St Eq
9 Sun	♍			E	Root	St Vo
10 Mon	♍			E	Root	
11 Tue	♍	○ 2a		E	Root	
12 Wed	♎ 12p			E/L	Root to 1p / Flower from 2p	St
13 Thu	♎			L	Flower to 12a	
14 Fri	♏ 2a	*Good Friday*	●☉	L/W		Eq Tr
15 Sat	♏	**Ag** 6a		W		
16 Sun	♐ 5p	*Easter*		W/H	Leaf from 1a to 4p / Fr 5p - 9p / 10p	St
17 Mon	♐	◡ 9a	▲	H	Leaf to 12a / Fruit from 1p	
18 Tue	♐			H	Fruit	St Eq
19 Wed	♑ 5a	☉-♈ ☾ 6a		H/E	Fruit - 4a / Root from 5a	
20 Thu	♑			E	Root	
21 Fri	♒ 10a	☍ 7p		E/L	Root to 9a / 10a - 1p / 10p	St
22 Sat	♒		●♆	L	Flower to 2p / Flower 7p	♄ St Eq
23 Sun	♓ 7a			L/W	Flower to 6a / Leaf from 7a (specially good from 2p)	St
24 Mon	♓		▲	W	Leaf to 12a (specially good to 7a)	St
25 Tue	♈ 10p			W/H		St
26 Wed	♈	● 8a	☿☍	H	Fruit from 2p to 12p	
27 Thu	♉ 12p	**Pg** 12p		H/E		St Vo
28 Fri	♉			E	Root from 1a	
29 Sat	♊ 10p			E/L	Root to 9p / 10p	
30 Sun	♊	⌒ 10a		L	Flower	

Northern Transplanting Time

12 1 2 3 4 5 6 7 8 9 10 11 12 1 2 3 4 5 6 7 8 9 10 11 12

Mercury ☿	Venus ♀	Mars ♂	Jupiter ♃	Saturn ♄	Uranus ♅	Neptune ♆	Pluto ♇
♈ 21 ♓	♓	♈	♍	♏	♓	♒	♐
(9 R)	(R 15 D)	12 ♉	(R)	(6 R)			(20 R)

NB: All zodiac symbols refer to astronomical constellations, not astrological signs (see p. 24)

Planetary aspects
(**Bold** = *visible to naked eye*)

April 2017

1	
2	☽☌♄ 11ᵃ
3	☽☍♇ 11ₚ
4	
5	♂△♇ 11ₚ
6	
7	☉☍♃ 6ₚ
8	☽☍♆ 1ᵃ
9	☽☍♀ 4ᵃ
10	☽☌♃ 7ₚ
11	☾☍⊕ 8ᵃ
12	☾☍☿ 4ᵃ
13	☾☌♂ 8ₚ
14	☉●⊕ 1ᵃ
15	
16	☾☌♄ 2ₚ
17	☉△♄ 9ᵃ
18	☾☌♇ 10ᵃ
19	
20	☉☌☿ 2ᵃ
21	
22	☾●♆ 4ₚ
23	☾☌♀ 6ₚ
24	☿△♄ 2ᵃ ☾☍♃ 11ₚ
25	☾☌⊕ 2ₚ ☾☌☿ 5ₚ
26	☿☊ 1ᵃ
27	
28	☽☌♂ 5ᵃ ☿☌⊕ 11ᵃ
29	☽☍♄ 5ₚ
30	

Mercury in Aries, initially supported by Mars, will bring warmth. When Mars moves into Taurus on April 12 cold will be added to the precipitation brought by other planets in Water constellations.

Northern Transplanting Time
April 3 4ᵃ to April 17 7ᵃ and
April 30 12ₚ to May 14
Southern Transplanting Time
March 21 to April 2 11ₚ and
April 17 11ᵃ to April 30 8ᵃ

The **soil warms up** on April 1.

Grafting of fruiting shrubs at Fruit times outside transplanting times.
    Grafting of flowering shrubs at Flower times outside transplanting times.

Control
Slugs from April 4 2ₚ to April 5 12ₚ.
Moths from April 23 7ᵃ to April 24 11ₚ.
Clothes and wax moths from April 26 2ₚ to April 26 11ₚ.

Southern hemisphere harvest time for seeds
Fruit seeds: April 5 12ₚ to April 8 8ₚ and at other Fruit times.
Flower seeds at Flower times, **Leaf seeds** at Leaf times, and **Root seeds** at Root times, always avoiding unfavourable times.

Biodynamic preparations: The preparations can be taken out of the earth, avoiding unfavourable times.

Planet (naked eye) visibility
Evening: Mercury (to 13th), Mars
All night: Jupiter, Saturn
Morning: Venus

Unfavourable time

May 2017

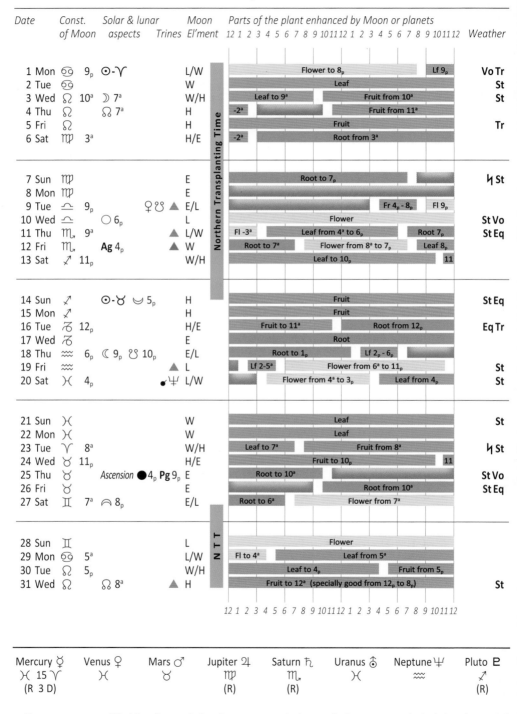

Date	Const. of Moon	Solar & lunar aspects	Moon Trines	El'ment	Parts of the plant enhanced by Moon or planets	Weather
1 Mon	♋ 9ₚ	☉-♈		L/W	Flower to 8ₚ — Lf 9ₚ	Vo Tr
2 Tue	♋			W	Leaf	St
3 Wed	♌ 10ᵃ	☽ 7ᵃ		W/H	Leaf to 9ᵃ / Fruit from 10ᵃ	St
4 Thu	♌	☍ 7ᵃ		H	-2ᵃ / Fruit from 11ᵃ	
5 Fri	♌			H	Fruit	Tr
6 Sat	♍ 3ᵃ			H/E	-2ᵃ / Root from 3ᵃ	
7 Sun	♍			E	Root to 7ₚ	♄ St
8 Mon	♍			E		
9 Tue	♎ 9ₚ	♀☍ ▲		E/L	Fr 4ₚ - 8ₚ — Fl 9ₚ	
10 Wed	♎	○ 6ₚ		L	Flower	St Vo
11 Thu	♏ 9ᵃ		▲	L/W	Fl -3ᵃ / Leaf from 4ᵃ to 6ₚ — Root 7ₚ	St Eq
12 Fri	♏	Ag 4ₚ	▲	W	Root to 7ᵃ / Flower from 8ᵃ to 7ₚ — Leaf 8ₚ	
13 Sat	♐ 11ₚ			W/H	Leaf to 10ₚ — 11	
14 Sun	♐	☉-♉ ☋ 5ₚ		H	Fruit	St Eq
15 Mon	♐			H	Fruit	
16 Tue	♑ 12ₚ			H/E	Fruit to 11ᵃ / Root from 12ₚ	Eq Tr
17 Wed	♑			E	Root	
18 Thu	♒ 6ₚ	☾ 9ₚ ☋ 10ₚ		E/L	Root to 1ₚ / Lf 2ₚ - 6ₚ	
19 Fri	♒		▲	L	Lf 2-5ᵃ / Flower from 6ᵃ to 11ₚ	St
20 Sat	♓ 4ₚ	☌♅		L/W	Flower from 4ᵃ to 3ₚ — Leaf from 4ₚ	St
21 Sun	♓			W	Leaf	St
22 Mon	♓			W	Leaf	
23 Tue	♈ 8ᵃ			W/H	Leaf to 7ᵃ / Fruit from 8ᵃ	♄ St
24 Wed	♉ 11ₚ			H/E	Fruit to 10ₚ — 11	
25 Thu	♉	Ascension ● 4ₚ Pg 9ₚ		E	Root to 10ᵃ	St Vo
26 Fri	♉			E	Root from 10ᵃ	St Eq
27 Sat	♊ 7ᵃ	⌒ 8ₚ		E/L	Root to 6ᵃ / Flower from 7ᵃ	
28 Sun	♊			L	Flower	St
29 Mon	♋ 5ᵃ			L/W	Fl to 4ᵃ / Leaf from 5ᵃ	
30 Tue	♌ 5ₚ			W/H	Leaf to 4ₚ / Fruit from 5ₚ	
31 Wed	♌	☍ 8ᵃ	▲	H	Fruit to 12ᵃ (specially good from 12ₚ to 8ₚ)	St

Northern Transplanting Time (7 Sun – 27 Sat) — *NTT* (28 Sun – 31 Wed)

Mercury ☿	Venus ♀	Mars ♂	Jupiter ♃	Saturn ♄	Uranus ♅	Neptune ♆	Pluto ♇
♓ 15 ♈	♓	♉	♍	♏	♓	♒	♐
(R 3 D)			(R)	(R)			(R)

NB: All zodiac symbols refer to astronomical constellations, not astrological signs (see p. 24)

Planetary aspects
(**Bold** = *visible to naked eye*)

1 ☽☌♇ 6ᵃ
2
3
4
5 ☽☌♆ 7ᵃ
6 ☽☌♀ 11ₚ

7 **☽☌♃ 7ₚ**
8 ☽☌☿ 4ₚ ☽☍☋ 5ₚ
9 ♀☋ 3ᵃ ☉△♇ 2ₚ
10 ☿☋ 1ᵃ
11 ☿△♄ 4ₚ
12 ♂△♃ 6ᵃ **☾☍♂ 7ₚ**
13 **☾☌♄ 7ₚ**

14
15 **☾☌♇ 4ₚ**
16
17
18
19 ♄△☊ 3ᵃ ♀☍♃ 10ᵃ
20 **☾•♆ 2ᵃ**

21
22 **☾☍♃ 6ᵃ ☾☌♀ 10ᵃ**
23 **☾☌☊ 3ᵃ ☾☌☿ 10ₚ**
24
25
26 ☽☌♂ 10ₚ
27 ☽☍♄ 1ᵃ

28 ☽☍♇ 2ₚ
29 ♂☍♄ 3ᵃ
30
31 ☿△♇ 8ᵃ

Planet (naked eye) visibility
Evening: Mars (to 22nd)
All night: Jupiter, Saturn
Morning: Venus

May is similar to April. Mercury is in Pisces, leaving this watery region on May 15 to Aries. In contrast to this Warmth influence, Mars in Taurus brings an underlying coolness.

Northern Transplanting Time
April 30 to May 14 3ₚ and
May 27 10ₚ to June 10
Southern Transplanting Time
May 14 7ₚ to May 27 6ₚ

Transplant **table potatoes** at Root times.
Transplant **seed potatoes** for 2018 from May 23 8ᵃ to May 24 10ₚ.

Hay should be cut between May 27 7ᵃ and May 29 4ᵃ, and at other Flower times.

Control:
Flies by burning fly papers in the cow barn at Flower times.
Chitinous insects, wheat weevil, Colorado beetle and varroa from May 24 11ₚ to May 27 6ᵃ.

Begin **queen bee** rearing (grafting or larval transfer, comb insertion, cell punching) at Flower times, especially when Moon is in Gemini (April 29 10ₚ to May 1 8ₚ, and May 27 7ᵃ to May 29 4ᵃ).

Biodynamic preparations: Cut birch, fill with yarrow and hang on May 19, 6ᵃ to 2ₚ.

▬▬▬ *Unfavourable time*   41

June 2017

All times in EDT

Date	*Const. of Moon*	*Solar & lunar aspects*	*Trines*	*Moon El'ment*	*Parts of the plant enhanced by Moon or planets* (12 1 2 3 4 5 6 7 8 9 10 11 12 1 2 3 4 5 6 7 8 9 10 11 12)	*Weather*
1 Thu	♌	☉-♉ ☽ 9^a	▲	H	Leaf from 1^a to 3_p; Fruit from 4_p	
2 Fri	♍ 8^a			H/E	Fruit to 7^a; Root from 8^a	♄ St
3 Sat	♍		▲	E	Root (specially good from 12^a to 3_p)	St Vo
4 Sun	♍	*Pentecost*		E	Root	St Eq Tr
5 Mon	♍			E	Root	Eq Tr
6 Tue	♎ 2^a			E/L	1^a; Flower from 2^a	St
7 Wed	♏ 3_p			L/W	Flower to 2_p; Leaf 3_p	
8 Thu	♏	**Ag** 6_p		W	Leaf to 9^a; Flower from 10^a to 9_p; 10_p	♄ St Vo
9 Fri	♏	○ 9^a		W	Leaf	
10 Sat	♐ 5^a	◡ 11_p		W/H	Lf to 4^a; Fruit from 5^a	
11 Sun	♐			H	Fruit	
12 Mon	♑ 5_p			H/E	Fruit to 4_p; Root from 5_p	
13 Tue	♑		▲	E	Root to 5_p (specially good from 12^a to 2_p)	♄ St Vo
14 Wed	♑	☋ 11_p	☿☊	E		St Eq
15 Thu	♒ 12^a			L	Flower from 7^a	
16 Fri	♒		☍♆	L	Flower to 6^a; Flower from 11^a to 11_p	
17 Sat	♓ 12^a	☾ 8^a		W	Leaf from 12^a	
18 Sun	♓			W	Leaf	
19 Mon	♈ 6_p			W/H	Leaf to 5_p; Fruit from 6_p	♄ St Tr
20 Tue	♈	☉-♊		H	Fruit	
21 Wed	♉ 9^a			H/E	Fruit to 8^a; Root from 9^a	St
22 Thu	♉			E	Root to 7_p	
23 Fri	♊ 6_p	**Pg** 7^a ● 11_p		E/L	Fruit 7_p	St
24 Sat	♊	◠ 7^a	▲	L	Fruit to 8^a; Flower from 9^a	
25 Sun	♋ 4_p			L/W	Flower	St
26 Mon	♋		▲	W	Flower to 5^a; Leaf from 6^a	♄ Tr
27 Tue	♌ 2^a	☊ 12_p	▲	W/H	1^a; Fruit 2^a to 9^a; Flower 4_p to 12^a	St
28 Wed	♌			H	Fruit from 12^a	
29 Thu	♍ 4_p			H/E	Fruit to 3_p; Root from 4_p	
30 Fri	♍	☽ 9_p		E	Root	

Northern Tr Time (1–10 June); N T T (24–30 June)

Mercury ☿	Venus ♀	Mars ♂	Jupiter ♃	Saturn ♄	Uranus ♅	Neptune ♆	Pluto ♇
♈ 3 ♉	♓ 5 ♈	♉	♍	♏	♓	♒	♐
21 ♊	29 ♉	4 ♊	(R 9 D)	(R)		(16 R)	(R)

NB: All zodiac symbols refer to astronomical constellations, not astrological signs (see p. 24)

Planetary aspects

*(**Bold** = visible to naked eye)*

1 ♀△♄ 11ᵃ ☽☌♆ 2ₚ
2
3 ♀☌⊕ 4ᵃ ☉△♃ 12ₚ ☽☌♃ 10ₚ

4
5 ☽☌⊕ 1ᵃ ☽☌♀ 5ᵃ
6
7 ☽☌☿ 11ₚ
8
9 ☾☌♄ 9ₚ
10 ☾☌♂ 4ₚ

11 ☾☌♇ 9ₚ
12
13 ☿△♃ 12ₚ
14 ☿♌ 5ₚ
15 ☉☌♄ 6ᵃ
16 ☾●♆ 9ᵃ
17

18 ☾☌♃ 1ₚ ☿☌♄ 3ₚ
19 ☾☌⊕ 2ₚ
20 ☾☌♀ 6ₚ
21 ☉☌☿ 10ᵃ
22
23 ☾☌♄ 9ᵃ
24 ☽☌☿ 4ᵃ ♀△♇ 5ᵃ ☽☌♂ 3ₚ

25 ☽☌♇ 1ᵃ
26 ♂△♆ 3ᵃ
27 ☿△♃ 8ₚ
28 ☿☌♂ 4ₚ ☽☌♆ 10ₚ
29 ☿☌♇ 9ₚ
30

The first few days of June will be similar to May. Then Mercury will bring cool influences from Taurus, while Venus, in Aries continues to bring Warmth. Mars in Gemini brings light influence, perhaps leading bees to make honeydew honey. Neither in May nor in June does Venus bring any Light infuence to enhance hay quality.

Northern Transplanting Time
May 27 to June 10 10ₚ and
June 24 9ᵃ to July 8
Southern Transplanting Time
June 11 2ᵃ to June 24 5ᵃ

Cut **hay** at Flower times.

Begin **queen bee** rearing June 15 7ᵃ to June 16 11ₚ and at other Flower times, avoiding unfavourable times.

Control:
Flies by burning fly papers in the cow barn at Flower times.
Mole crickets ash from June 7 3ₚ to June 10 4ᵃ.
Grasshoppers from June 23 6ₚ to June 25 3ₚ.

June

Planet (naked eye) visibility
Evening:
All night: Jupiter, Saturn
Morning: Venus

Unfavourable time   43

July 2017

All times in EDT

Date	Const. of Moon	Solar & lunar aspects	Moon Trines	El'ment	Parts of the plant enhanced by Moon or planets (12 1 2 3 4 5 6 7 8 9 10 11 12 1 2 3 4 5 6 7 8 9 10 11 12)	Weather
1 Sat	♍	☉-♊		E	Root	♄ Eq Tr
2 Sun	♍			E	Root	
3 Mon	♎ 9ᵃ			E/L	Root to 8ᵃ / Flower from 9ᵃ	
4 Tue	♏ 9ₚ			L/W	Flower to 8ₚ / Leaf 9ₚ	St
5 Wed	♏		▲	W	Leaf to 8ᵃ / Fl from 9ᵃ (sp good from 5ₚ)	♄ St Eq
6 Thu	♏	Ag 1ᵃ		W	Fl to 4ᵃ / Leaf from 5ᵃ	
7 Fri	♐ 11ᵃ			W/H	Leaf to 10ᵃ / Fruit from 11ᵃ	St Vo Tr
8 Sat	♐	☽ 7ᵃ		H	Fruit	♄ St
9 Sun	♑ 11ₚ	○ 1ᵃ		H/E	Fruit to 10ₚ 11	
10 Mon	♑			E	Root	St Eq
11 Tue	♑			E	Root to 10ₚ	
12 Wed	♒ 6ᵃ	☌ 1ᵃ		E/L	Flower from 6ᵃ	
13 Thu	♒		•♅	L	Flower to 12ₚ / Flower from 5ₚ	
14 Fri	♓ 5ᵃ			L/W	Fl to 4ᵃ / Leaf from 5ᵃ	
15 Sat	♓			W	Leaf	St Vo
16 Sun	♓	☾ 3ₚ		W	Leaf to 12ᵃ	
17 Mon	♈ 1ᵃ			H	Fruit from 1ᵃ	♄ St Eq
18 Tue	♉ 5ₚ		▲	H/E	Fr to 4ᵃ / Root from 5ᵃ (sp good from 9ᵃ to 7ₚ)	♄ St
19 Wed	♉			E	Root	St Eq
20 Thu	♉	☉-♋		E	Root to 12ᵃ	St
21 Fri	♊ 4ᵃ	Pg 1ₚ ☌ 6ₚ		E/L		St Eq Vo
22 Sat	♊			L		
23 Sun	♋ 1ᵃ	● 6ᵃ	☿♋	W	Leaf from 1ₚ	
24 Mon	♌ 12ₚ	☌ 9ₚ		W/H	Leaf to 11ᵃ / Fr 12 - 4ₚ	
25 Tue	♌		•☿	H	1-3ᵃ / Fruit from 8ᵃ	
26 Wed	♌			H	Fruit to 11ₚ	
27 Thu	♍ 12ᵃ			E	Root from 12ᵃ	St Tr
28 Fri	♍			E	Root	St Vo
29 Sat	♍			E	Root	St
30 Sun	♎ 4ₚ	☽ 11ᵃ		E/L	Root to 3ₚ / Flower from 4ₚ	St Eq
31 Mon	♎			L	Flower	

(12 1 2 3 4 5 6 7 8 9 10 11 12 1 2 3 4 5 6 7 8 9 10 11 12)

Northern Transplanting Time — NTT

Mercury ☿	Venus ♀	Mars ♂	Jupiter ♃	Saturn ♄	Uranus ♅	Neptune ♆	Pluto ♇
♊ 4 ♋	♉	♊	♍	♏	♓	♒	♐
17 ♌	31 ♊	16 ♋		(R)		(R)	(R)

NB: All zodiac symbols refer to astronomical constellations, not astrological signs (see p. 24)

Planetary aspects
(**Bold** = *visible to naked eye*)

July 2017

1 ☽♂♃ 6ᵃ

2 ♂♂♇ 8ᵃ ☽♂♅ 9ᵃ
3

4

5 ☽♂♀ 2ᵃ ☉△♆ 9ₚ
6

7 ☽♂♄ 1ᵃ
8

9 ☾♂♇ 2ᵃ ☾♂♂ 12ₚ
10 ☉♂♇ 1ᵃ ☾♂♀ 7ₚ
11

12

13 ☾•♆ 2ₚ
14

15 ☾♂♃ 11ₚ

16 ☾♂♅ 10ₚ
17

18 ♀△♃ 4ₚ
19 ☿△♄ 3ₚ
20 ☾♂♀ 8ᵃ ☾♂♄ 4ₚ
21

22 ☾♂♇ 9ᵃ

23 ☿☍ 1ᵃ ☽♂♂ 8ᵃ
24 ♀♂♄ 11ᵃ ☿△♅ 1ₚ
25 ☽•☿ 5ᵃ
26 ☽♂♆ 7ᵃ ☉♂♂ 9ₚ
27

28 ☽♂♃ 7ₚ
29 ☽♂♅ 6ₚ

30

31

The planetary influences of the first half of the year give the impression that Mercury and Venus have conspired to avoid any continuous weather influences. On July 16 Mercury moves into Leo, hopefully heralding Warmth and Fruit influences.

Northern Transplanting Time
June 24 to July 8 5ᵃ and
July 21 8ₚ to Aug 4
Southern Transplanting Time
July 8 9ᵃ to July 21 4ₚ

Late hay cut at Flower times.

Summer harvest for seeds:
Flower plants: Harvest at Flower times, specially in the first half of the month.
   Fruit plants from July 24 12ₚ to July 26 11ₚ, avoiding unfavourable times, or at other Fruit times.
   Harvest **leaf plants** at Leaf times.
   Harvest **root plants** at Root times, especially July 18 5ᵃ to July 20 11ₚ.

Control
Flies: burn fly papers in the cow barn at Flower times.
Slugs: burn between July 23 1ᵃ and July 24 11ᵃ. Spray leaf plants and the soil with horn silica early in the morning during Leaf times.

Biodynamic preparations
Pick **valerian** early in the morning at Flower times. Quickly press out the sap without diluting with water or laying the plants in water. Diluted sap does not keep as long. Search out valerian plants well beforehand to ensure a speedy harvest.
   Cut **oak,** fill it with ground oak bark and put it in the earth on July 2 between 5ᵃ and 12ₚ.

Planet (naked eye) visibility
Evening: Jupiter
All night: Saturn
Morning: Venus

Unfavourable time   45

All times in EDT

Date	Const. of Moon	Solar & lunar aspects	Moon Trines	El'ment	Parts of the plant enhanced by Moon or planets 12 1 2 3 4 5 6 7 8 9 10 11 12 1 2 3 4 5 6 7 8 9 10 11 12	Weather

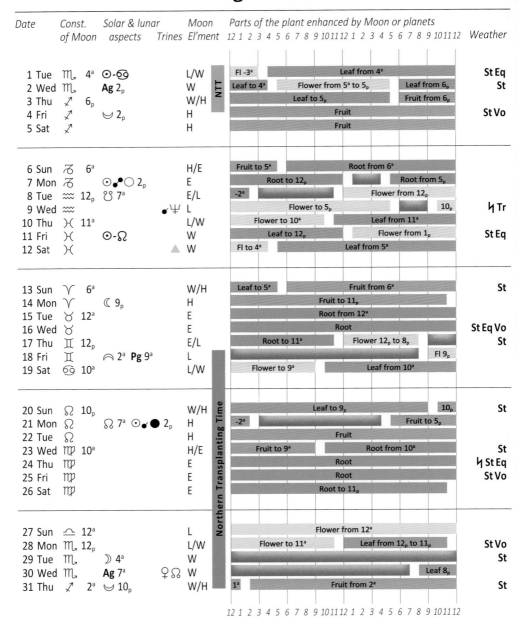

1 Tue ♏ 4ᵃ	☉-♋		L/W	Fl -3ᵃ / Leaf from 4ᵃ	St Eq
2 Wed ♏ Ag 2ₚ			W	Leaf to 4ᵃ / Flower from 5ᵃ to 5ₚ / Leaf from 6ₚ	St
3 Thu ♐ 6ₚ			W/H	Leaf to 5ₚ / Fruit from 6ₚ	
4 Fri ♐ ☽ 2ₚ			H	Fruit	St Vo
5 Sat ♐			H	Fruit	

6 Sun ♑ 6ᵃ			H/E	Fruit to 5ᵃ / Root from 6ᵃ	
7 Mon ♑ ☉●○ 2ₚ			E	Root to 12ₚ / Root from 5ₚ	
8 Tue ♒ 12ₚ ☍ 7ᵃ			E/L	-2ᵃ / Flower from 12ₚ	
9 Wed ♒	●♅		L	Flower to 5ₚ / 10ₚ	♄ Tr
10 Thu ♓ 11ᵃ			L/W	Flower to 10ᵃ / Leaf from 11ᵃ	
11 Fri ♓ ☉-♌			W	Leaf to 12ₚ / Flower from 1ₚ	St Eq
12 Sat ♓		▲	W	Fl to 4ᵃ / Leaf from 5ᵃ	

13 Sun ♈ 6ᵃ			W/H	Leaf to 5ᵃ / Fruit from 6ᵃ	St
14 Mon ♈ ☾ 9ₚ			H	Fruit to 11ₚ	
15 Tue ♉ 12ᵃ			E	Root from 12ᵃ	
16 Wed ♉			E	Root	St Eq Vo
17 Thu ♊ 12ₚ			E/L	Root to 11ᵃ / Flower 12ₚ to 8ₚ	St
18 Fri ♊ ⌢ 2ᵃ Pg 9ᵃ			L	Fl 9ₚ	
19 Sat ♋ 10ᵃ			L/W	Flower to 9ᵃ / Leaf from 10ᵃ	

20 Sun ♌ 10ₚ			W/H	Leaf to 9ₚ / 10ₚ	St
21 Mon ♌ ☍ 7ᵃ ☉●● 2ₚ			H	-2ᵃ / Fruit to 5ₚ	
22 Tue ♌			H	Fruit	
23 Wed ♍ 10ᵃ			H/E	Fruit to 9ᵃ / Root from 10ᵃ	St
24 Thu ♍			E	Root	♄ St Eq
25 Fri ♍			E	Root	St Vo
26 Sat ♍			E	Root to 11ₚ	

27 Sun ♎ 12ᵃ			L	Flower from 12ᵃ	
28 Mon ♏ 12ₚ			L/W	Flower to 11ᵃ / Leaf from 12ₚ to 11ₚ	St Vo
29 Tue ♏ ☽ 4ᵃ			W		St
30 Wed ♏ Ag 7ᵃ	♀☍		W	Leaf 8ₚ	
31 Thu ♐ 2ᵃ ☾ 10ₚ			W/H	1ᵃ / Fruit from 2ᵃ	St

12 1 2 3 4 5 6 7 8 9 10 11 12 1 2 3 4 5 6 7 8 9 10 11 12

(Side labels: NTT; Northern Transplanting Time)

Mercury ☿	Venus ♀	Mars ♂	Jupiter ♃	Saturn ♄	Uranus ♅	Neptune ♆	Pluto ♇
♌	♊	♋	♍	♏	♓	♒	♐
(12 R)	24 ♋	18 ♌		(R 25 D)	(3 R)	(R)	(R)

NB: All zodiac symbols refer to astronomical constellations, not astrological signs (see p. 24)

August 2017

1	
2	
3	☽☌♄ 4ᵃ
4	☽☍♀ 5ᵃ
5	☽☌♇ 8ᵃ
6	
7	☽☍♂ 7ᵃ
8	
9	☾☍☿ 3ₚ ☾●♅ 7ₚ
10	
11	
12	♀△♅ 1ᵃ ☾☍♃ 11ᵃ
13	☾☌☊ 4ᵃ ☉△♄ 5ₚ
14	
15	♀☌♇ 7ᵃ
16	☾☍♄ 10ₚ
17	
18	☾☍♇ 5ₚ
19	☾☌♀ 1ᵃ
20	
21	☾☌♂ 1ᵃ ☉△☊ 2ᵃ
22	☽☌☿ 6ᵃ ♂△♄ 9ᵃ ☽☍♅ 3ₚ
23	
24	
25	☽☌♃ 12ₚ
26	☽☍☊ 2ᵃ ☉☌☿ 5ₚ
27	
28	
29	
30	♀☊ 7ᵃ ☽☌♀ 11ᵃ
31	

Mercury is in Leo for the whole month, and Venus (until Aug 23) in the Light constellation of Gemini. The Sun (from Aug 11) and Mars (from Aug 18) move into Leo, hopefully bringing Warmth for a good harvest.

Northern Transplanting Time
July 21 to Aug 4 12ₚ and
Aug 18 4ᵃ to Aug 31 8ₚ
Southern Transplanting Time
Aug 4 4ₚ to Aug 17 11ₚ

Harvest **seeds of fruit plants** and **grain** to be used for seed from Aug 20 10ₚ to Aug 23 9ᵃ, and at other Fruit times, avoiding unfavourable times.

Immediately after harvest, sow catch crops like lupins, phacelia, mustard or wild flax.

Seeds for leaf plants: harvest at Leaf times, specially in the first half of the month.

Seeds for flower plants: at Flower times, specially in the second week of August.

Burn **fly papers** in the cow barn from Aug 17 12ₚ to Aug 19 9ᵃ, and at other Flower times.

Ants in the house: burn when the Moon is in Leo, Aug 20 10ₚ to Aug 23 9ᵃ, avoiding unfavourable times.

Biodynamic preparations: Cut **maple** and fill with dandelion and put it in the earth on Aug 15 from 4ᵃ to 9ᵃ.

Cut **yarrow** in the mornings at Flower times from Aug 11. The blossoms should show some seed formation.

Aug

Planet (naked eye) visibility
Evening: Jupiter
All night: Saturn
Morning: Venus

September 2017

Date	Const. of Moon	Solar & lunar aspects	Trines	Moon El'ment	Parts of the plant enhanced by Moon or planets	Weather

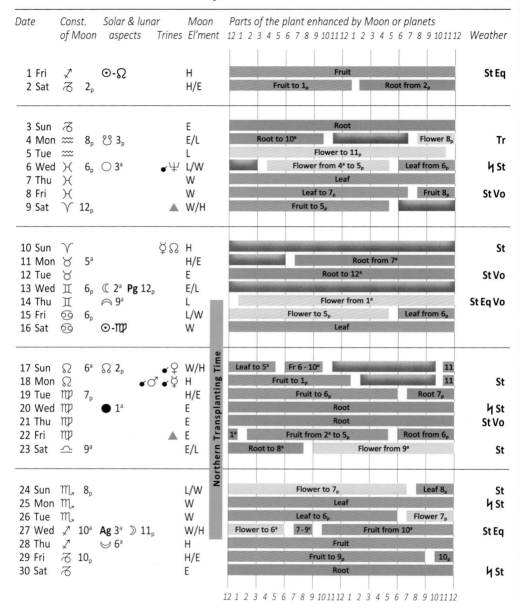

Date	Const.	aspects	Trines	El'ment	Weather
1 Fri	♐	☉-♌		H	St Eq
2 Sat	♑	2ₚ		H/E	
3 Sun	♑			E	
4 Mon	♒	8ₚ ☊ 3ₚ		E/L	Tr
5 Tue	♒			L	
6 Wed	♓	6ₚ ○ 3ª	♂♆	L/W	♄ St
7 Thu	♓			W	
8 Fri	♓			W	St Vo
9 Sat	♈	12ₚ	▲	W/H	
10 Sun	♈		☿☊	H	St
11 Mon	♉	5ª		H/E	St Vo
12 Tue	♉			E	
13 Wed	♊	6ₚ ☽ 2ª **Pg** 12ₚ		E/L	St Eq Vo
14 Thu	♊		⌒ 9ª	L	
15 Fri	♋	6ₚ		L/W	
16 Sat	♋	☉-♍		W	
17 Sun	♌	6ª ☊ 2ₚ	♀	W/H	
18 Mon	♌		♂ ♀	H	St
19 Tue	♍	7ₚ		H/E	
20 Wed	♍	● 1ª		E	♄ St
21 Thu	♍			E	St Vo
22 Fri	♍		▲	E	
23 Sat	♎	9ª		E/L	St
24 Sun	♏	8ₚ		L/W	St
25 Mon	♏			W	♄ St
26 Tue	♏			W	
27 Wed	♐	10ª **Ag** 3ª ☽ 11ₚ		W/H	St Eq
28 Thu	♐	⌣ 6ª		H	
29 Fri	♑	10ₚ		H/E	
30 Sat	♑			E	♄ St

(Northern Transplanting Time — label along vertical axis)

Bar chart part labels:

- 1 Fri: Fruit
- 2 Sat: Fruit to 1ₚ / Root from 2ₚ
- 3 Sun: Root
- 4 Mon: Root to 10ª / Flower 8ₚ
- 5 Tue: Flower to 11ₚ
- 6 Wed: Flower from 4ª to 5ₚ / Leaf from 6ₚ
- 7 Thu: Leaf
- 8 Fri: Leaf to 7ₚ / Fruit 8ₚ
- 9 Sat: Fruit to 5ₚ
- 10 Sun: (Root)
- 11 Mon: Root from 7ª
- 12 Tue: Root to 12ª
- 13 Wed: Flower from 1ª
- 14 Thu: Flower to 5ₚ / Leaf from 6ₚ
- 15 Fri: Flower to 5ₚ / Leaf from 6ₚ
- 16 Sat: Leaf
- 17 Sun: Leaf to 5ª / Fr 6 - 10ª / 11
- 18 Mon: Fruit to 1ₚ / 11
- 19 Tue: Fruit to 6ₚ / Root 7ₚ
- 20 Wed: Root
- 21 Thu: Root
- 22 Fri: 1ª / Fruit from 2ª to 5ₚ / Root from 6ₚ
- 23 Sat: Root to 8ª / Flower from 9ª
- 24 Sun: Flower to 7ₚ / Leaf 8ₚ
- 25 Mon: Leaf
- 26 Tue: Leaf to 6ₚ / Flower 7ₚ
- 27 Wed: Flower to 6ª / 7 - 9ª / Fruit from 10ª
- 28 Thu: Fruit
- 29 Fri: Fruit to 9ₚ / 10ₚ
- 30 Sat: Root

12 1 2 3 4 5 6 7 8 9 10 11 12 1 2 3 4 5 6 7 8 9 10 11 12

Mercury ☿	Venus ♀	Mars ♂	Jupiter ♃	Saturn ♄	Uranus ♅	Neptune ♆	Pluto ♇
♌ 26 ♍	♋	♌	♍	♏	♓	♒	♐
(R 5 D)	10 ♌				(R)	(R)	(R 28 D)

NB: All zodiac symbols refer to astronomical constellations, not astrological signs (see p. 24)

Planetary aspects
 (**Bold** = _visible to naked eye_)

1 ☽☌♇ 2ₚ
2 ♂△⚷ 8ᵃ

3 ☿☌♂ 6ᵃ ☽☍♀ 12ₚ
4 ☽☍☿ 11ₚ
5 ☽☍♂ 1ᵃ ☉☍♆ 1ᵃ
6 ☽●♆ 1ᵃ
7
8
9 ☾☍♃ 1ᵃ ☉△♇ 7ᵃ ☾☌⚷ 9ᵃ

10 ☿♌ 5ₚ
11
12 ♀△♄ 9ₚ
13 ☾☍♄ 4ᵃ
14 ☾☍♇ 11ₚ
15
16 ☿☌♂ 3ₚ

17 ☾●♀ 9ₚ ♀△⚷ 11ₚ
18 ☾●♂ 4ₚ ☾●☿ 7ₚ ☾☍♆ 11ₚ
19 ☿☍♆ 11ₚ
20
21
22 ☽☌♃ 6ᵃ ☽☍⚷ 9ᵃ ☿△♇ 2ₚ
23

24 ♂☍♆ 4ₚ
25
26 ☽☌♄ 8ₚ
27 ♃☍⚷ 11ₚ
28 ☽☌♇ 10ₚ
29 ♀☍♆ 8ₚ
30

Planet (naked eye) visibility
Evening: Jupiter
All night: Saturn
Morning: Mercury (4th to 28th), Venus, Mars (from 11th)

The Warmth of August should last until Sep 25, as Mercury, Mars and Venus (from Sep 10) are in Leo, and should dominate the other planets in Water and Earth constellations.

Northern Transplanting Time
Sep 14 11ᵃ to Sep 28 4ᵃ
Southern Transplanting Time
Sep 1 1ᵃ to Sep 14 7ᵃ and
Sep 28 8ᵃ to Oct 11

The times recommended for the **fruit harvest** are those in which the Moon is in Aries or Sagittarius (Aug 31 2ᵃ to Sep 2 1ₚ, Sep 9 12ₚ to 5ₚ, and Sep 27 10ᵃ to Sep 29 9ₚ) or other Fruit times.
  The harvest of **root crops** is always best undertaken at Root times. Storage trials of onions, carrots, beetroot and potatoes have demonstrated this time and again.

Good times for **sowing winter grain** are when the Moon is in Leo or Sagittarius (Aug 31 2ᵃ to Sep 2 1ₚ, Sep 17 6ᵃ to Sep 19 6ₚ, and Sep 27 10ᵃ to Sep 29 9ₚ) avoiding unfavourable times, and at other Fruit times.
  Rye can if necessary also be sown at Root times with all subsequent cultivations being carried out at Fruit times.

Control slugs by burning between Sep 15 6ₚ and Sep 17 5ᵃ.

Biodynamic preparations: Cut larch and fill with chamomile and put it in the earth between Sep 19 9ₚ and Sep 20 5ᵃ.

 Unfavourable time   49

October 2017

All times in EDT

Date	Const. of Moon		Solar & lunar aspects	Trines	Moon El'ment	Parts of the plant enhanced by Moon or planets (12 1 2 3 4 5 6 7 8 9 10 11 12 1 2 3 4 5 6 7 8 9 10 11 12)	Weather
1 Sun	♑		☉-♍ ☋ 10_p	▲	E	Root to 7^a · Fruit from 8^a to 6_p · ▒	
2 Mon	♒	5^a			E/L	▒ · $2 - 4^a$ · Flower from 5^a	
3 Tue	♒			•♆ ▲	L	Flower to 6^a · ▒ · Fruit $11^a - 6_p$ · Flower 7_p	Tr
4 Wed	♓	2^a			L/W	1^a · Leaf from 2^a	
5 Thu	♓		○ 3_p		W	Leaf	Tr
6 Fri	♈	8_p			W/H	Leaf to 7_p · Fruit 8_p	
7 Sat	♈				H	Fruit	♄ St Eq
8 Sun	♉	12_p			H/E	Fruit to 11^a · ▒	♄ St
9 Mon	♉		**Pg** 2^a		E	▒ · Root from 2_p	St Vo
10 Tue	♊	11_p			E/L	Root to 10_p · 11	St Eq Vo
11 Wed	♊		⌒ 2_p		L	Flower	St Eq
12 Thu	♋	11_p	☾ 8^a		L/W	Flower to 10_p · 11	
13 Fri	♋				W	Leaf	
14 Sat	♌	12_p	☊ 6_p		W/H	Leaf to 11^a · $12\text{-}2_p$ · ▒ · Fr 9_p	St
15 Sun	♌				H	Fruit	St Eq
16 Mon	♌				H	Fruit	
17 Tue	♍	2^a			H/E	1^a · Root from 2^a	
18 Wed	♍			☿☋	E	1^a · ▒	
19 Thu	♍		● 3_p		E	▒ · Root from 12_p	St Eq
20 Fri	♎	5_p			E/L	Root to 4_p · Flower from 5_p	Vo Tr
21 Sat	♎				L	Flower	
22 Sun	♏	4^a			L/W	Fl -3^a · Leaf from 4^a	St
23 Mon	♏				W	Leaf to 11_p	
24 Tue	♐	6_p	**Ag** 11_p	▲	W/H	Flower from 12^a to 5_p · Fruit from 6_p	Tr
25 Wed	♐		◡ 2_p		H	Fruit	
26 Thu	♐				H	Fruit	
27 Fri	♑	7^a	☽ 6_p		H/E	Fruit to 6^a · Root from 7^a	
28 Sat	♑				E	Root to 11_p	St Eq Vo Tr
29 Sun	♒	2_p	☋ 3^a		E/L	▒ · Rt 7^a to 1_p · Flower from 2_p	St
30 Mon	♒			•♆	L	Flower to 3_p · ▒ · Flower 8_p	
31 Tue	♓	12_p			L/W	Flower to 11^a · Leaf from 12_p	

Northern Transplanting Time (October 11 – October 24)

Mercury ☿	Venus ♀	Mars ♂	Jupiter ♃	Saturn ♄	Uranus ⛢	Neptune ♆	Pluto ♇
♍	♌	♌	♍	♏	♓	♒	♐
23 ♎	9 ♍	12 ♍			(R)	(R)	

NB: All zodiac symbols refer to astronomical constellations, not astrological signs *(see p. 24)*

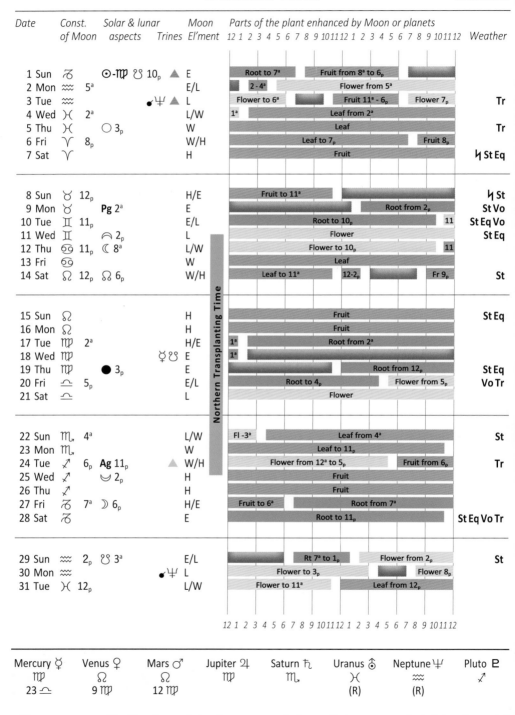

Planetary aspects
*(**Bold** = visible to naked eye)*

1	♂△♇ 8ₚ
2	
3	☽●♂ 9ᵃ ♀△♇ 3ₚ ☽☍♀ 6ₚ ☽☍♂ 8ₚ
4	
5	☽☍☿ 10ᵃ ♀♂♂ 1ₚ
6	☾♂♁ 3ₚ ☾♂♃ 7ₚ
7	
8	☉♂☿ 5ₚ
9	
10	☾☍♄ 12ₚ
11	
12	☾☍♇ 4ᵃ
13	
14	
15	☿☍♁ 4ᵃ
16	☾☍♆ 5ᵃ
17	☾♂♂ 7ᵃ ☾♂♀ 10ₚ
18	☿♂♃ 5ᵃ ☿⅋⅋ 11ₚ
19	☉☍♁ 1ₚ ☾♂♁ 3ₚ
20	☽♂♃ 2ᵃ ☽♂☿ 7ᵃ
21	
22	
23	
24	☽♂♄ 8ᵃ ☿△♆ 12ₚ
25	
26	☽♂♇ 7ᵃ ☉♂♃ 2ₚ
27	
28	
29	
30	☽●♆ 6ₚ
31	

October should be ideal for harvesting root crops. Sun, Mercury, Jupiter, Venus (from Oct 9) and Mars (from Oct 12) are all in the cool Earth constellation of Virgo. Other planets support with Water, Light and Warmth. Hopefully, the cool influences will not bring frost.

Northern Transplanting Time
Oct 11 4ₚ to Oct 25 12ₚ
Southern Transplanting Time
Sep 28 to Oct 11 12ₚ and
Oct 25 4ₚ to Nov 7

Store fruit at any Fruit or Flower time outside transplanting time.

Harvest seeds of root plants at Root times, **seeds for leaf plants** at Leaf times, and **seeds for flower plants** at Flower times.

All **cleared ground** should be treated with compost and sprayed with barrel preparation, and ploughed ready for winter.

Control slugs by burning between Oct 12 11ₚ and Oct 14 11ᵃ.

Planet (naked eye) visibility
Evening: Jupiter (to 2nd), Saturn
All night:
Morning: Venus, Mars

Unfavourable time

Oct

November 2017

Date	Const. of Moon	Solar & lunar aspects	Moon Trines	El'ment	Parts of the plant enhanced by Moon or planets	Weather

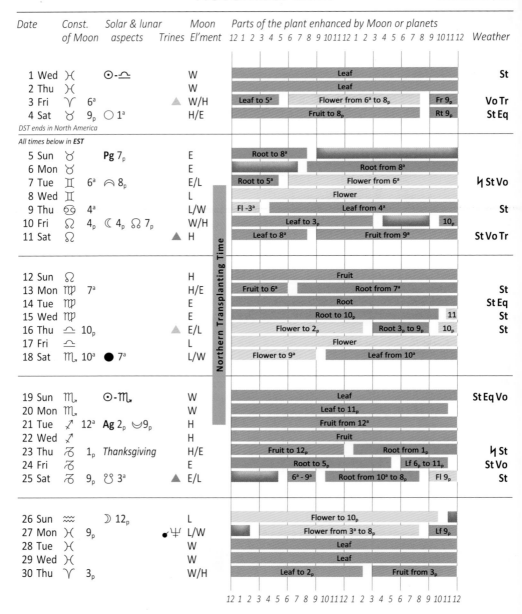

Parts of the plant enhanced by Moon or planets — scale: 12 1 2 3 4 5 6 7 8 9 10 11 12 1 2 3 4 5 6 7 8 9 10 11 12

1 Wed ♓ — ☉-♎ — W — Leaf — St
2 Thu ♓ — W — Leaf
3 Fri ♈ 6ᵃ — ▲ W/H — Leaf to 5ᵃ / Flower from 6ᵃ to 8ₚ / Fr 9ₚ — Vo Tr
4 Sat ♉ 9ₚ ○ 1ᵃ — H/E — Fruit to 8ₚ / Rt 9ₚ — St Eq

DST ends in North America

All times below in EST

5 Sun ♉ — Pg 7ₚ — E — Root to 8ᵃ
6 Mon ♉ — E — Root from 8ᵃ
7 Tue ♊ 6ᵃ ♑ 8ₚ — E/L — Root to 5ᵃ / Flower from 6ᵃ — ♄ St Vo
8 Wed ♊ — L — Flower
9 Thu ♋ 4ᵃ — L/W — Fl -3ᵃ / Leaf from 4ᵃ — St
10 Fri ♌ 4ₚ ☾ 4ₚ ☊ 7ₚ — W/H — Leaf to 3ₚ / 10ₚ
11 Sat ♌ — ▲ H — Leaf to 8ᵃ / Fruit from 9ᵃ — St Vo Tr

12 Sun ♌ — H — Fruit
13 Mon ♍ 7ᵃ — H/E — Fruit to 6ᵃ / Root from 7ᵃ — St
14 Tue ♍ — E — Root — St Eq
15 Wed ♍ — E — Root to 10ₚ / 11 — St
16 Thu ♎ 10ₚ — ▲ E/L — Flower to 2ₚ / Root 3ₚ to 9ₚ / 10ₚ — St
17 Fri ♎ — L — Flower
18 Sat ♏ 10ᵃ ● 7ᵃ — L/W — Flower to 9ᵃ / Leaf from 10ᵃ

19 Sun ♏ — ☉-♏ — W — Leaf — St Eq Vo
20 Mon ♏ — W — Leaf to 11ₚ
21 Tue ♐ 12ᵃ Ag 2ₚ ☋ 9ₚ — H — Fruit from 12ᵃ
22 Wed ♐ — H — Fruit
23 Thu ♑ 1ₚ *Thanksgiving* — H/E — Fruit to 12ₚ / Root from 1ₚ — ♄ St
24 Fri ♑ — E — Root to 5ₚ / Lf 6ₚ to 11ₚ — St Vo
25 Sat ♑ 9ₚ ☊ 3ᵃ — ▲ E/L — 6ᵃ - 9ᵃ / Root from 10ᵃ to 8ₚ / Fl 9ₚ — St

26 Sun ♒ — ☽ 12ₚ — L — Flower to 10ₚ
27 Mon ♓ 9ₚ — •♆ L/W — Flower from 3ᵃ to 8ₚ / Lf 9ₚ
28 Tue ♓ — W — Leaf
29 Wed ♓ — W — Leaf
30 Thu ♈ 3ₚ — W/H — Leaf to 2ₚ / Fruit from 3ₚ

Northern Transplanting Time

scale: 12 1 2 3 4 5 6 7 8 9 10 11 12 1 2 3 4 5 6 7 8 9 10 11 12

Mercury ☿	Venus ♀	Mars ♂	Jupiter ♃	Saturn ♄	Uranus ♅	Neptune ♆	Pluto ♇
♎ 4 ♏	♍ 15 ♎	♍	♍	♏	♓	♒	♐
30 ♐	29 ♏		24 ♎		(R)	(R 22 D)	

NB: All zodiac symbols refer to astronomical constellations, not astrological signs (see p. 24)

Planetary aspects
(**Bold** = *visible to naked eye*)

1 ☽☍♂ 2p
2 ☽☍♀ 9p ☽☌♁ 11p
3 ☽☍♃ 2p ☉△♆ 3p
4 ♀☌♁ 1a

5 ☾☍☿ 4a
6 ☾☌♄ 10p
7
8 ☾☍♇ 10a
9
10
11 ♄△♁ 4a

12 ☾☍♆ 8a
13 ♀☌♃ 3a
14 ☾☌♂ 10p
15 ☾☍♁ 7p
16 ♀△♆ 10a ☾☌♃ 7p
17 ☾☌♀ 3a
18

19
20 ☽☌☿ 6a ☽☌♄ 7p
21
22 ☽☌♇ 2p
23
24
25 ☿△♁ 6a

26
27 ☽☌♆ 1a
28 ☿☌♄ 2a
29
30 ☽☍♂ 6a ☽☌♁ 7a

Planet (naked eye) visibility
Evening: Saturn
All night:
Morning: Venus, Mars, Jupiter (from 8th)

The Sun moves into Libra on November 1, and into Scorpio on Nov 19. The planets move alternately in Water and Earth constellations, with occaional Light and Warmth times, giving a colourful mix.

Northern Transplanting Time
Nov 7 10p to Nov 21 7p
Southern Transplanting Time
Oct 25 to Nov 7 6p and
Nov 21 11p to Dec 5

The Flower times in Transplanting Time are ideal for **planting flower bulbs,** showing vigorous growth and vivid colours. The remaining Flower times should only be considered as back up, as bulbs planted on those times will not flower so freely.

If not already completed in October, all organic waste materials should be gathered and made into a **compost.** Applying the biodynamic preparations to the compost will ensure a rapid transformation and good fungal development. An application of barrel preparation will also help the composting process.

Fruit and forest trees will also benefit at this time from a spraying of horn manure and/or barrel preparation when being transplanted.

Best times for **cutting Advent greenery** and **Christmas trees** for transporting are Flower times outside Transplanting Time (Nov 3 6a to 8p, Nov 7 6a to 10p and Nov 25 9p to Nov 27 8p).

Burn **fly papers** in cow barn at Flower times, Nov 7 6a to Nov 9 3a and Nov 16 10p to Nov 18 9a.

Biodynamic preparations: Put birch and yarrow into the ground betweeen Nov 3 11p and Nov 4 7a.

Nov

Unfavourable time

December 2017

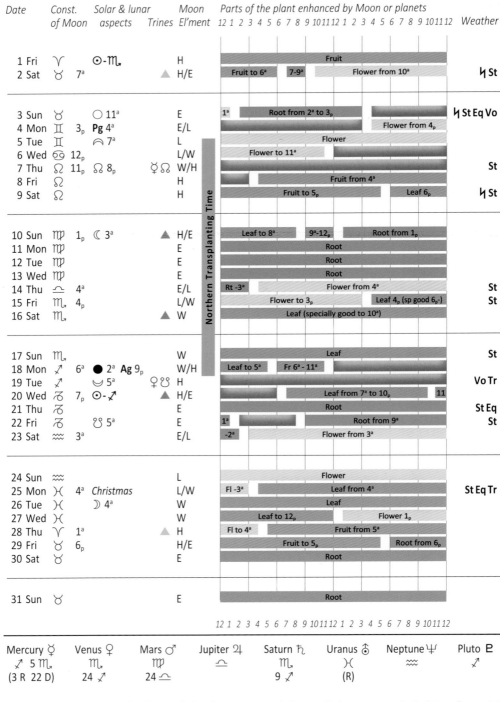

Date	Const. of Moon	Solar & lunar aspects	Moon Trines	El'ment	Parts of the plant enhanced by Moon or planets	Weather

1 Fri ♈ ☉-♏ H — Fruit — ♄ St
2 Sat ♉ 7ᵃ ▲ H/E Fruit to 6ᵃ / 7-9ᵃ / Flower from 10ᵃ

3 Sun ♉ ○ 11ᵃ E 1ᵃ Root from 2ᵃ to 3ₚ ♄ St Eq Vo
4 Mon ♊ 3ₚ **Pg** 4ᵃ E/L Flower from 4ₚ
5 Tue ♊ ⌒ 7ᵃ L Flower
6 Wed ♋ 12ₚ L/W Flower to 11ᵃ
7 Thu ♌ 11ₚ ♌♋ 8ₚ ☿♌ W/H Fruit from 4ᵃ — St
8 Fri ♌ H
9 Sat ♌ H Fruit to 5ₚ / Leaf 6ₚ ♄ St

10 Sun ♍ 1ₚ ☾ 3ᵃ ▲ H/E Leaf to 8ᵃ / 9ᵃ-12ₚ / Root from 1ₚ
11 Mon ♍ E Root
12 Tue ♍ E Root
13 Wed ♍ E Root
14 Thu ♎ 4ᵃ E/L Rt -3ᵃ / Flower from 4ᵃ St
15 Fri ♏ 4ₚ L/W Flower to 3ₚ / Leaf 4ₚ (sp good 6ₚ-) St
16 Sat ♏ ▲ W Leaf (specially good to 10ᵃ)

17 Sun ♏ W Leaf St
18 Mon ♐ 6ᵃ ● 2ᵃ **Ag** 9ₚ W/H Leaf to 5ᵃ / Fr 6ᵃ - 11ᵃ
19 Tue ♐ ☋ 5ᵃ ♀♌ H Vo Tr
20 Wed ♑ 7ₚ ☉-♐ ▲ H/E Leaf from 7ᵃ to 10ₚ / 11 St Eq
21 Thu ♑ E Root
22 Fri ♑ ♌ 5ᵃ E 1ᵃ Root from 9ᵃ St
23 Sat ♒ 3ᵃ E/L -2ᵃ Flower from 3ᵃ

24 Sun ♒ L Flower
25 Mon ♓ 4ᵃ *Christmas* L/W Fl -3ᵃ / Leaf from 4ᵃ St Eq Tr
26 Tue ♓ ☽ 4ᵃ W Leaf
27 Wed ♓ W Leaf to 12ₚ / Flower 1ₚ
28 Thu ♈ 1ᵃ ▲ H Fl to 4ᵃ / Fruit from 5ᵃ
29 Fri ♉ 6ₚ H/E Fruit to 5ₚ / Root from 6ₚ
30 Sat ♉ E Root

31 Sun ♉ E Root

12 1 2 3 4 5 6 7 8 9 10 11 12 1 2 3 4 5 6 7 8 9 10 11 12

Northern Transplanting Time

Mercury ☿	Venus ♀	Mars ♂	Jupiter ♃	Saturn ♄	Uranus ♅	Neptune ♆	Pluto ♇
♐ 5 ♏	♏	♍	♎	♏	♓	♒	♐
(3 R 22 D)	24 ♐	24 ♎		9 ♐	(R)		

NB: All zodiac symbols refer to astronomical constellations, not astrological signs (see p. 24)

Planetary aspects
   (**Bold** = *visible to naked eye*)

1	♂☍♁ 5^a ☽☍♃ 10^a
2	☽☍♀ 8_p ♃△♆ 10_p
3	
4	☾☌♄ 1_p ☾☍☿ 2_p
5	☾☍♇ 8_p
6	☿☌♄ 7^a
7	☿☊ 3_p
8	
9	☾☍♆ 3_p
10	☿△♁ 5^a
11	
12	☉☌☿ 9_p ☾☍♁ 11_p
13	☾☌♂ 2_p
14	☾☌♃ 12_p
15	☿☌♀ 9^a
16	☉△♁ 6^a
17	☾☌☿ 4^a ☾☌♀ 2_p
18	☽☌♄ 8^a
19	♀☊ 7_p ☽☌♇ 10_p
20	♀△♁ 6_p
21	☉☌♄ 4_p
22	
23	
24	☽☌♆ 9^a
25	♀☌♄ 1_p
26	
27	☽☌♁ 4_p
28	♂△♆ 1^a ☽☍♂ 11_p
29	☽☍♃ 5^a
30	
31	☽☍☿ 8^a

Planet (naked eye) visibility
Evening: Saturn (to 4th)
All night:
Morning: Venus (to 9th), Mars, Jupiter

December continues with a colourful mix. Mercury and Venus interchange between warm Sagittarius and watery Scorpio. Mars in Virgo brings cold, and Jupiter strengthens Light influences. With Saturn moving from a Water to a Warmth constellation, Uranus bringing Water influences, Neptune in Light and Pluto in Warmth, it is unlikely to be a white Christmas.

Northern Transplanting Time
Dec 5 9^a to Dec 19 3^a
Southern Transplanting Time
Nov 21 to Dec 5 5^a and
Dec 19 7^a to Jan 1

The transplanting time is good for **pruning trees and hedges.** Fruit trees should be pruned at Fruit or Flower times.

Best times for cutting **Advent greenery** and **Christmas trees** are at Flower times to ensure lasting fragrance.

Burn feathers or skins of **warm blooded pests** from Dec 2 7^a to Dec 3 3_p. Ensure the fire is glowing hot (don't use grilling charcoal). Lay dry feathers or skins on the glowing embers. After they have cooled, collect the light grey ash and grind for an hour with a pestle and mortar. This increases the efficacy and the ash can be potentised later. *The burning and grinding should be completed by Dec 3 3_p.*

Southern hemisphere:
Harvest time for seeds (avoid unfavourable times):
Leaf seeds: Leaf times.
Fruit seeds: Fruit times, preferably with Moon in Leo (Dec 8 4^a to Dec 9 5_p).
Root seeds: Root times.
Flower seeds: Flower times.

Control slugs Dec 6 12_p to Dec 7 10_p.

We wish all our readers a blessed Advent and Christmastide and the best of health for the New Year of 2018

▬▬▬ *Unfavourable time*

Sowing times for trees and shrubs

March 24: Apple, Beech, Alder

April 21: Apple, Beech, Ash, Sweet chestnut

June 15 (after 7[a]): Alder, Larch, Lime tree, Elm, Thuja, Juniper, Plum, Hornbeam

June 30: Pear, Birch, Lime tree, Robinia, Willow

July 10: Ash, Cedar, Fir, Spruce, Hazel

July 24: Birch, Pear, Lime tree, Robinia, Willow, Thuja, Juniper, Plum, Hornbeam

Sep 9: Ash, Cedar, Fir, Spruce, Hazel

Sep 20: Alder, Larch, Lime tree, Elm

Oct 19 (after 12[p]): Ash, Spruce, Hazel, Fir, Cedar

Dec 1: Yew, Oak, Cherry, Horse chestnut (buckeye), Sweet chestnut, Spruce, Fir, Thuja, Juniper, Hornbeam

The above dates refer to sowing times of the seeds. They are not times for transplanting already existing plants.

The dates given are based on planetary aspects, which create particularly favourable growing conditions for the species in question. For trees and shrubs not mentioned above, sow at an appropriate time of the Moon's position in the zodiac, depending on the part of the tree or shrub to be enhanced. Avoid unfavourable times.

Felling times for timber

June 3: Maple, Apple, Copper beech, Sweet chestnut, Walnut, Spruce, Hornbeam, Pine, Fir, Thuja, Cedar, Plum, Plum, Alder, Larch, Lime tree, Elm, Birch, Pear, Robinia, Willow

June 28: Alder, Larch, Lime tree, Elm

July 18: Birch, Pear, Larch, Lime tree, Robinia, Willow

Oct 24: Alder, Larch, Lime tree, Elm

Nov 3: Ash, Spruce, Hazel, Fir, Cedar

Dec 21: Birch, Pear, Larch, Lime tree, Robinia, Willow

Those trees which are not listed should be felled during November and December at Flower times during the descending Moon period (transplanting time). Avoid unfavourable times.

Types of crop

Flower plants

artichoke
broccoli
flower bulbs
flowering ornamental shrubs
flowers
flowery herbs
rose
sunflower

Leaf plants

asparagus
Brussels sprouts
cabbage
cauliflower
celery
chard
chicory (endive)
Chinese cabbage (pe-tsai)
corn salad (lamb's lettuce)
crisphead (iceberg) lettuce
curly kale (green cabbage)
endive (chicory)
finocchio (Florence fennel)
green cabbage (curly kale)
iceberg (crisphead) lettuce
kohlrabi
lamb's lettuce (corn salad)
leaf herbs
leek
lettuce
pe-tsai (Chinese cabbage)
red cabbage
rhubarb
shallots
spinach

Root plants

beetroot
black (Spanish) salsify
carrot
celeriac
garlic
horseradish
Jerusalem artichoke
parsnip
potato
radish
red radish
root tubers
Spanish (black) salsify

Fruit plants

aubergine (eggplant)
bush bean
courgette (zucchini)
cucumber
eggplant (aubergine)
grains
lentil
maize
melon
paprika
pea
pumpkin (squash)
runner bean
soya
squash (pumpkin)
tomato
zucchini (courgette)

The care of bees

A colony of bees lives in its hive closed off from the outside world. For extra protection against harmful influences, the inside of the hive is sealed with propolis. The link with the wider surroundings is made by the bees that fly in and out of the hive.

To make good use of cosmic rhythms, the beekeeper needs to create the right conditions in much the same way as the gardener or farmer does with the plants. The gardener works the soil and in so doing allows cosmic forces to penetrate it via the air. These forces can then be taken up and used by the plants until the soil is next moved.

When the beekeeper opens up the hive, the sealing layer of propolis is broken. This creates a disturbance, as a result of which cosmic forces can enter and influence the life of the hive until the next intervention by the beekeeper. By this means the beekeeper can directly mediate cosmic forces to the bees.

It is not insignificant which forces of the universe are brought into play when the the hive is opened. The beekeeper can consciously intervene by choosing days for working with the hive that will help the colony to develop and build up its food reserves. The bees will then reward the beekeeper by providing a portion of their harvest in the form of honey.

Earth-Root times can be selected for opening the hive if the bees need to do more building. Light-Flower times encourage brood activity and colony development. Warmth-Fruit times stimulate the collection of nectar. Water-Leaf times are unsuitable for working in the hive or for the removal and processing of honey.

Since the late 1970s the varroa mite has affected virtually every bee colony in Europe. Following a number of comparative trials we recommend burning and making an ash of the varroa mite in the usual way. After dynamising it for one hour, the ash should be put in a salt-cellar and sprinkled lightly between the combs. The ash should be made and sprinkled when the Sun and Moon are in Taurus (May/June).

Feeding bees in preparation for winter

The herbal teas recommended as supplements in the feeding of bees prior to winter are all plants that have proved their value over many years. Yarrow, chamomile, dandelion and valerian are made by pouring boiling water over the flowers, allowing them to brew for fifteen minutes and then straining them. Stinging nettle, horsetail and oak bark are placed in cold water, brought slowly to the boil and simmered for fifteen minutes. Three grams (1 tablespoon) of each dried herb and half a litre (½ quart) of the prepared teas is enough to produce 100 litres (25 gal) of liquid feed. This is a particularly important treatment in years when there are large amounts of honeydew.

Fungal problems

The function of fungus in nature is to break down dying organic materials. It appears amongst our crops when unripe manure compost or uncomposted animal by-products such as horn and bone meal are used but also when seeds are harvested during unfavourable constellations: according to Steiner, 'When Moon forces are working too strongly on the Earth ...'

Tea can be made from horsetail (*Equisetum arvense*) and sprayed on to the soil where affected plants are growing. This draws the fungal level back down into the ground where it belongs.

The plants can be strengthened by spraying stinging nettle tea on the leaves. This will promote good assimilation, stimulate the flow of sap and help fungal diseases to disappear.

Biodynamic preparation plants

Pick **dandelions** in the morning at Flower times as soon as they are open and while the centre of the flowers are still tightly packed.

Pick **yarrow** at Fruit times when the Sun is in Leo (around the middle of August).

Pick **chamomile** at Flower times just before midsummer. If they are harvested too late, seeds will begin to form and there are often grubs in the hollow heads.

Collect **stinging nettles** when the first flowers are opening, usually around midsummer. Harvest the whole plants without roots at Flower times.

Pick **valerian** at Flower times around midsummer.

All the flowers (except valerian) should be laid out on paper and dried in the shade.

Collect **oak bark** at Root times. The pithy material below the bark should not be used.

Biodynamic preparations: putting birch and yarrow into the ground

Moon diagrams

The diagrams overleaf show for each month the daily position (evenings GMT) of the Moon against the stars and other planets. For viewing in the southern hemisphere, turn the diagrams upside down.

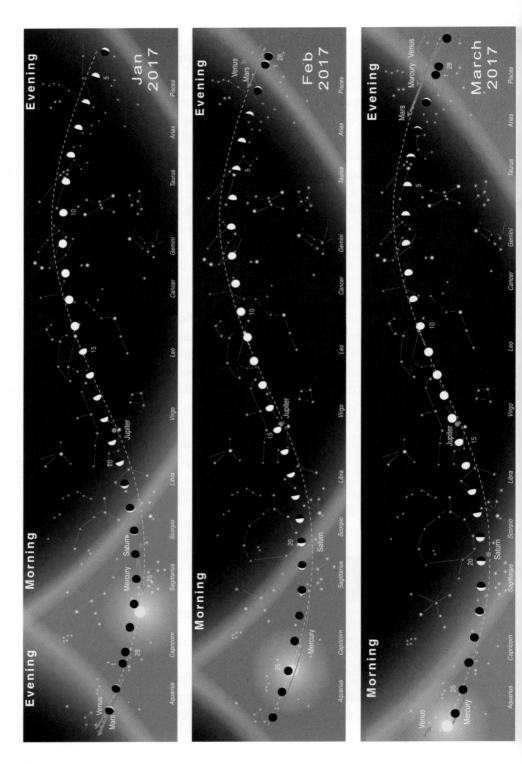

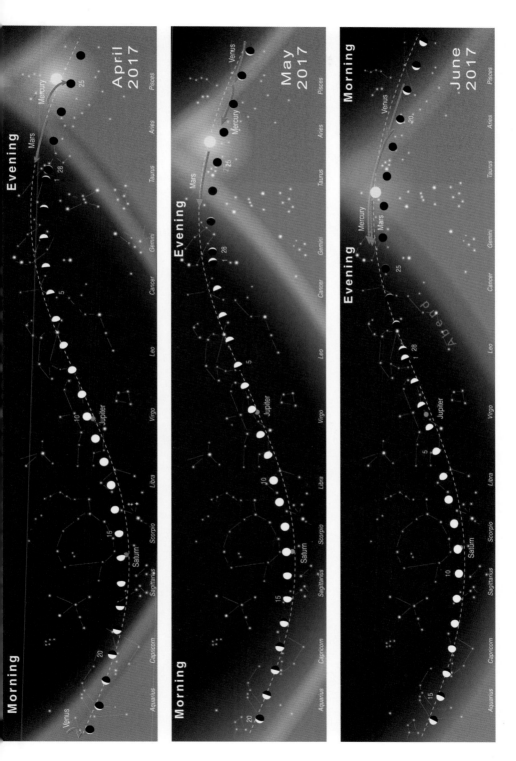

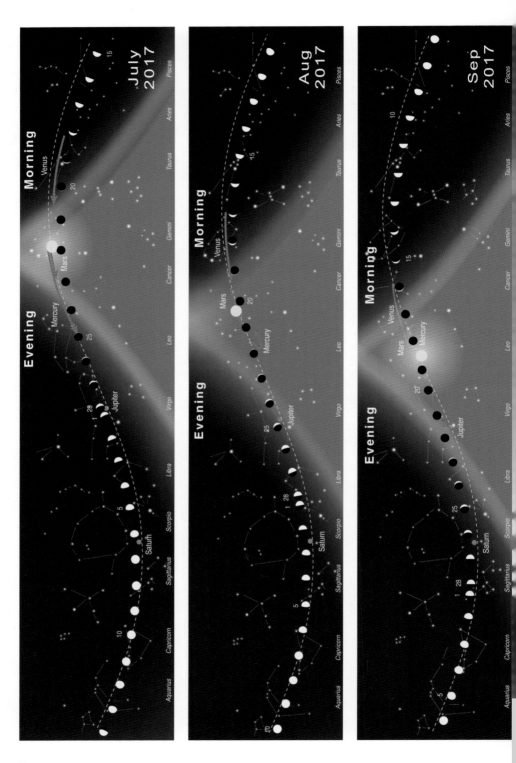

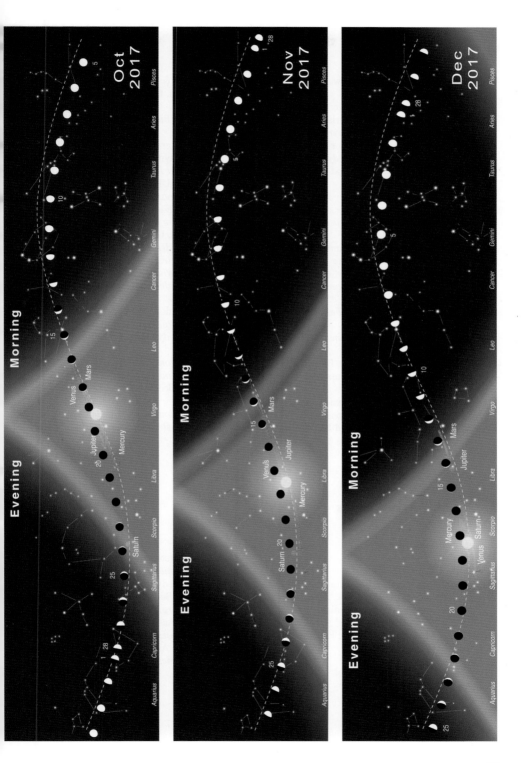

63

Further Reading

Colquhoun, Margaret and Axel Ewald, *New Eyes for Plants,* Hawthorn

Karlsson, Britt and Per, *Biodynamic, Organic and Natural Winemaking,* Floris

Klett, Manfred, *Principles of Biodynamic Spray and Compost Preparations,* Floris

Koepf, H.H., *Koepf's Practical Biodynamics: Soil, Compost, Sprays and Food Quality,* Floris

Kranich, Ernst Michael, *Planetary Influences upon Plants,* Biodynamic Association, USA

Lepetit, Antoine, *What's so Special About Biodynamic Wine?* Floris

Masson, Pierre, *A Biodynamic Manual,* Floris

Morrow, Joel, *Vegetable Gardening for Organic and Biodynamic Growers,* Lindisfarne

Osthaus, Karl-Ernst, *The Biodynamic Farm,* Floris

Pfeiffer, Ehrenfried, *The Earth's Face,* Lanthorn

—, *Pfeiffer's Introduction to Biodynamics,* Floris

—, *Weeds and What They Tell Us,* Floris

—, & Michael Maltas, *The Biodynamic Orchard Book,* Floris

Philbrick, John and Helen, *Gardening for Health and Nutrition,* Anthroposophic, USA

Philbrick, Helen & Gregg, Richard B., *Companion Plants and How to Use Them,* Floris

Sattler, Friedrich & Eckard von Wistinghausen, *Growing Biodynamic Crops,* Floris

Schilthuis, Willy, *Biodynamic Agriculture,* Floris

Steiner, Rudolf, *Agriculture (A Course of Eight Lectures),* Biodynamic Association, USA

—, *Agriculture: An Introductory Reader,* Steiner Press, UK

—, *What is Biodynamics? A Way to Heal and Revitalize the Earth,* SteinerBooks, USA

Storl, Wolf, *Culture and Horticulture,* North Atlantic Books, USA

Thun, Maria, *Gardening for Life,* Hawthorn

—, *The Biodynamic Year,* Temple Lodge

Thun, Matthias, *When Wine Tastes Best: A Biodynamic Calendar for Wine Drinkers,* (annual) Floris

von Keyserlink, Adelbert Count, *The Birth of a New Agriculture,* Temple Lodge

—, *Developing Biodynamic Agriculture,* Temple Lodge

Waldin, Monty, *Monty Waldin's Best Biodynamic Wines,* Floris

Weiler, Michael, *Bees and Honey, from Flower to Jar,* Floris

Wright, Hilary, *Biodynamic Gardening for Health and Taste,* Floris

Biodynamic Associations

Demeter International
www.demeter.net

Australia:
Bio-Dynamic Research Institute
www.demeter.org.au
Biodynamic Agriculture Australia
www.biodynamics.net.au

Canada: Society for Bio-Dynamic Farming & Gardening in Ontario
biodynamics.on.ca

India: Bio-Dynamic Association of India (BDAI)
www.biodynamics.in

New Zealand:
Biodynamic Association
www.biodynamic.org.nz

South Africa: Biodynamic Agricultural Association of Southern Africa
www.bdaasa.org.za

UK: Biodynamic Association
www.biodynamic.org.uk

USA: Biodynamic Association
www.biodynamics.com

Hunting Pennsylvania Turkeys

TOM FEGELY

Published By

B & T Outdoor Enterprises
P.O. Box 986
Cherryville, PA 18035

Additional books can be ordered at:
www.fegelyoutdoors.com
or through B&T Outdoor Enterprises

ISBN 0-9643278-1-3

DEDICATION

To Colton, Mason, Skylar, Nick and Aaron,
all of whom hold potential for being hunters
and the next generation's stewards
of the great outdoors.

TABLE OF CONTENTS

SECTION I: THE EVOLUTION OF PENNSYLVANIA TURKEY HUNTING

SECTION II: THE BIOLOGY OF THE BIRD

SECTION III: GETTING READY: Guns, Bows, Garb and Gear

SECTION IV: WHERE TO GO: Turkeys from Forest to Farm

SECTION V: BEFORE THE HUNT PREP: Safety, Patterning and Choosing Calls

ACKNOWLEDGMENTS

A project such as this may have only one author's name but the people who have made contributions to it run into the dozens.

Over the past three decades I've learned most of what I know about turkey hunting from a long list of "turkey pros," many of whom serve as pro-staffers for game call, firearms or camouflage garb companies. Joining them afield offered me the opportunities to experience first-hand what others hear at seminars or watch vicariously on video. Quotes and tidbits gleaned from stories I wrote about them over the years provided invaluable reference material.

The list include the likes of some nationally and locally familiar names including: Ben Rogers Lee, Rob Keck, Carl Brown, Brad Harris, Eddie Salter, Matt Morrett, Will Primos, Joe Drake, Bob Clark, Glenn Lindaman, Harold Knight, Dave Hale, Ronnie Strickland, Paul Butski, Billy Macoy, Ronnie Johnson, Barry Haydt, Lovett Williams, J. Wayne Fears, Jim Casada, Bo Pittman, Preston Pitman, Larry Norton, Jerry Peterson, Shirley Grenoble, Dick Kirby, Ernie Calandrelli, Tes Jolly, Ronnie Jolly, Harry Boyer, Brian Post, Dave Streb, Don Jacobs, Dale Butler, Greg Caldwell, Doug Marquardt, Rob Poorman, Bill Jordan, Terry Rohm, Frank Piper, Greg Neumann, Tom Neumann, Scott Basehore, Toby Bridges, Jody Hugill, Ray Eye, Mike Wadell, David Blanton, Chuck Jones, Kelly Cooper, Walter Parrott, Wayne Gendron, Pete Clare, Steve Lecorchick and …. others whose names do not immediately come to mind.

For making available old records and reports, then scouring Pennsylvania Game Commission photo files for scenes of early turkey hunting management, I thank Jerry Feaser, Hal Korber and Joe Kosack. Invaluable help in the biological studies was provided by PGC turkey biologist Mary Jo Casalena and Bob Eriksen, Pennsylvania's biologist for the National Wild Turkey Federation. Add to the thank you list Bedford County WCO Tim Flanigan for his text and excellent photography and artist Mike Watson for his sketches.

You will also note many references to Arnie Hayden, taken from a taped interview with the PGC biologist just before his death in 2001. The late Jerry Wunz, another PGC biologist, also made important biological contributions during his career. He passed away in 2004.

My gratitude also goes to Tammy Sapp and Matt Lindler of the NWTF staff and to former Pennsylvanian and longtime friend Rob Keck, now the federation's CEO, for writing the Foreword and to Don Heckman and Jerry Zimmeran for their help.

Although there's insufficient space to mention their names here, I'm indebted to the 50-plus hunters, outdoor writers and Pennsylvania NWTF Chapter members who agreed to write articles on assigned subjects regarding turkey hunting and related activities. Their works, scattered throughout the book, appropriately fill the banners titled "Another View." Kudos also to fellow hunters Gary Visgaitis, formerly of U.S.A. Today graphics dept. for his design expertise and Tom Turner of Seaber Turner Assoc. for his helpful advice.

Finally, without my wife Betty Lou and her computer skills and attention to detail, this book would probably be another year in coming – if at all. We shared many midnight drills in our separate offices to meet our deadline and countless hours on the phone and email lines. Many gremlins can sneak into the production of a book, especially a self-published work such as this, and she found and corrected them.

Without her, this book truly might not have been possible.

T.D.F.

FOREWORD

First, thanks must go to my mentor, outdoor hero and good friend Tom Fegely, who I met many years ago and have had the joy of partnering with on gobbler hunts from Pennsylvania's Pine Creek to South Texas.

I thank him for what he meant to me in my formative years as an outdoor communicator in telling the story of hunters, hunting and what they've meant to the success of the greatest conservation model in the history of the world.

As hunters, we too often speak only to ourselves. In Tom's long career as an outdoor writer and photographer, he's told the hunter's story in print, from the podium and through the airwaves. His strong voice, great camera presence, good humor and tremendous grasp of biological, social, and strategic aspects of the hunt have provided a positive influence for hunters (and some non-hunters) way beyond the Keystone State's borders.

This was acknowledged in 2002 when Tom received the prestigious National Wild Turkey Federation's Communicator of the Year Award. But his achievements go far beyond that. His awards and recognitions read like a who's who in the outdoor industry. That claim is underscored by Tom's induction into the Pennsylvania Turkey Hunter's Hall of Fame in 1991.

Now he's made yet another contribution to the hunter's world with the publication of "A Guide to Hunting Pennsylvania Turkeys." It follows his successful "A Guide to Hunting Pennsylvania Whitetails" first published in 1994 and which is sold out of three printings.

His new book – with the help of his wife Betty Lou, also an avid turkey hunter -- is the culmination of many years of studying turkeys, photographing them, hunting them and painstakingly researching the biology of the bird.

Coincidentally, Tom sold his first article in 1968, the same year the state's spring gobbler season made its debut. Up until then it was fall hunting only. The wild turkey's restoration also spawned a large "camo clan" of hunters who no longer need to travel very far from their front doors to hear the yodel of a gobbler.

Tom also played a role in the battle to shut down the old game farm via his media outlets. He brought to his audiences the Pennsylvania Chapter's desire to close the farm and switch the funding to trap-and-transfer. Today we are, indeed, enjoying the best years of turkey hunting in Pennsylvania.

Tom has now put his experiences into a powerful book that will provide solid advice in helping you become a more successful and more knowledgeable hunter, along with numerous Pennsylvania "turkey tales" to boot. You'll not only learn about the Eastern wild turkey and its biology and history but how to hunt it and savor the memories. He includes his personal thoughts as well as tapping the minds of more than 50 other Pennsylvanians known for their turkey hunting talents or having voluntarily and valuably served the state chapter over the years.

No matter if you're a newcomer or seasoned veteran, I can guarantee that his richly-illustrated, all-Pennsylvania turkey hunting guide will bring hours of good reading at home and benefits afield.

Tom Fegely has, indeed, answered the call.

Rob Keck
CEO, National Wild Turkey Federation

PREFACE

As this is being written I can gaze out my office window at an array of yellow spicebush and golden beech leaves scattered across my 35 acre hillside woodlot.

It's October and I should be out bowhunting; although I admit to some selfish comfort that none of my friends have called to tell me they've seen many deer or tied their tags to a big one. With a little bit of luck another week will find me in one of the treestands dotting the property awaiting one of the pre-rut bucks I watched grow throughout the summer come into view.

Right now I'm inundated with turkeys; not in my woods but in my mind.

Simply, I am in the process of meeting deadline for the publication of the book you are now holding. It's a comforting feeling as my wife Betty Lou and I have been putting in 12 hour work days writing, researching, editing, choosing photos and writing cutlines to complete the job before our computers get the fits and sign off.

It's been that way for nearly two months. Turkeys are so ingrained in my thought processes that I see or hear them everywhere, even flying through my dreams at night.

In 1994, when I published and later sold out "A Guide to Hunting Pennsylvania Whitetails," I planned a quick follow-up with "A Guide to Hunting Pennsylvania Turkeys." But near daily newspaper and magazine deadlines threw a glitch into our plans. Now, 10 years later, it is completed.

I do not speak as a wild turkey authority, even though my degrees are in education and the biological sciences. However, I have been blessed with an insatiable curiosity about these enthralling birds and over my near-30 years as a full-time outdoor communicator have become friends and hunting partners with some of the best known names in the turkey hunting field.

Accompanying them afield, seeing what they did in particular situations, listening to their calls and studying their woodsmanship was like having a series of one-on-one seminars that provided me with much of the material you'll find in this book.

In the Acknowledgments I listed as many names as I could instantly recall who have contributed to my turkey education. Rest assured anyone I've missed will let me know about it posthaste. Nevertheless, thanks to all of them.

Perhaps the biggest overall lesson has been that no two hunters approach the challenge of getting close to a gobbler in the same way. You'll note that as you read this book. Each includes his or her own little twists and, in some cases, may use a technique completely contrary to someone else's when toms refuse to gobble or you just can't seem to get any to come near.

You will also read time and again about playing things safe; for safety must always be the primary concern of the hunter, as must honorable ethics. Secondly, you will note an emphasis throughout on not moving when a bird is in sight. I've known hunters, me included, who had to learn things the hard way and didn't really grasp the advice of others until I did it myself a few times.

In addition to my well-known mentors I asked more than 50 Pennsylvania hunters to respond to questions I posed to them regarding everything from locating birds and choosing a choke tube to picking the right call and being part of a turkey hunting family. I think you'll find their thoughts —which come under the heading "Another View" -- to be informative, innovative and intriguing.

I have also attempted to balance the text with variety, offering both hard-core, how-to advice tendered with softer, story-telling pieces. Add to that Pennsylvania turkey history, biology, restoration, dressing for the hunt, optics, shotgun patterning, hunting spring and fall, making videos and photos, taxidermy, recipes, ethics, laws and profiles of the NWTF and its Pennsylvania chapter.

I trust I have dealt with everything a novice or seasoned turkey hunter may want to know. Oh, yes. Add recruiting women and kids to the 41-chapter work; the future of hunting and retaining our cherished outdoors lifestyles.

It must also be noted that, with minor exception, all photography in this book was taken in Pennsylvania. I've carried a camera afield on nearly every hunt for the past 30 years or so, long before the mandatory fluorescent orange hat law was in effect. Most shots were taken on the scene following the harvest of a bird or in another non-hunting location. In most situations where orange was mandatory, however, the subjects were wearing the required safety color.

I hope you enjoy this guide and that the insights it offers will aid in putting a tag on a fan-tailed tom while making some indelible memories along the way.

Tom Fegely
Walnutport, PA
Oct. 22, 2004

SECTION I

The Evolution of Pennsylvania Turkey Hunting

A woman holds a turkey she shot with her double-barreled shotgun in this circa 1945 photograph.

CHAPTER 1

THROUGH THE YEARS
A Diary of Pennsylvania Turkeys and Turkey Hunting

"Rough, rocky and rewarding" may be a fitting slogan honoring Pennsylvania's long-traveled road in managing and restoring Meleagris gallopavo silvestris – the Eastern wild turkey – over the past century and more.

While newcomers to the spring and fall quests may remember the happenings of a mere few years ago, senior hunters recall the late 1960s when the initial spring turkey season was held and management began to be accepted as a science, not merely as "let's do something and see what happens" experiment.

Following is a historical diary of the status of turkeys over the past couple centuries. It puts into perspective the role of Pennsylvania hunters and biologists of the past as compared to today's contingent of turkey enthusiasts.

After scanning this diary I think you'll agree that for the wild turkey and those of us who dream about and pursue them, the good, old days are now.

• **1600s era** – Indians cherished, hunted and even tamed the abundant wild turkeys throughout the state.

Pennsylvania artist Jack Paluh captured this scene of the state's earliest turkey hunters in his "Native Hunters" work. www.jackpaluh.com

- **1683** – "Turkeys of the wood, I have had of 40- and 50-pound weight" wrote William Penn, who obviously exaggerated his claim yet recognized that many turkeys were about at that time ….. hunting permitted on all lands under Penn's Charter.
- **1700s** – Not many, if any, recreational hunters ….. turkeys largely a food source ….. attitude was: "If game disappeared from an area it meant you had to hunt somewhere else."
- **Early 1800s** – "They (turkeys) are becoming less numerous in every portion of the United States ….. even in those parts where they were very abundant 30 years ago." (John James Audubon).
- **1850-1860** – "Turkeys are extirpated in the New England states and New York," wrote ornithologist W.E. Clyde Todd.
- **Mid 1800s into early 1900s** – Turkey strongholds composed of Pennsylvania's ridge-and-valley region across the southcentral counties.
- **1873** – Turkey hunting outlawed from Jan. 1 to Oct. 1 ….. $25 fine for violators ….. Sunday hunting banned.
- **1888** – Pennsylvania ornithologist B.H. Warren wrote: "This noble gamebird, although rapidly becoming extirpated, is still found in small numbers in the wooded, thinly-populated and uncultivated districts of this Commonwealth."
- **1895** – Pennsylvania Game Commission is created ….. turkeys beginning to repopulate in some large, clearcut tracts ….. heavy understory of saplings and briars make them more difficult to hunt.
- **1897** – Daily limit for turkeys set at two.
- **1900** – Presumption by hunters and biologists that only a few thousand wild turkeys remained in the state's 45,333 square miles ….. not even tracks were easy to find in some places ….. no effort to curb hunting pressure.
- **1904** – PGC Executive Secretary Joseph Kalbfus claims state's wild turkey flocks beginning to return but adds: "They are being hunted too early, the season should be shorter, the use of dogs in hunting turkeys should be prohibited and hunting shouldn't start before sunrise" ….. state's General Assembly did not agree but Kalbfus maintained his stance ….. several years later legislators approved the dog ban, made blinds and turkey calls illegal, banned the sale of turkey meat and outlawed night hunting of birds on their roosts.
- **1905** – Daily limit reduced to one per day with a season limit of

By the turn of the century turkeys were wiped out in many places due to uncontrolled timber harvests.

four ..... season ran six weeks in October and November ..... deforestation taking a toll on habitat.

- **1909** – Use of turkey calls forbidden.
- **1913** – Estimated turkey hunter numbers set at 150,000 ..... more than 1,000 turkeys shot ..... use of turkey calls made illegal.
- **1914-1915** – Gov. John Tener closes statewide turkey season ..... considered a wise and timely move.
- **1915** – Turkeys first trapped and transferred in the state.
- **1922** – Game Commission raises turkeys and stocks birds raised out-of-state ..... PGC Executive Secretary Seth Gordon directs commissioners to purchase 100 Mexican wild turkeys at a cost not to exceed $10 each.
- **1923-1928** – Experimental trapping of wild turkeys on refuge areas ..... birds transferred to areas suffering low populations.
- **1926** – Turkey season again closed.
- **1928** – State ornithologist George Miksch Sutton writes: "Once fairly common throughout the state ..... brought to the verge of extermination by forest fires, excessive hunting and encroaches of civilization ..... some mingling has occurred between wild birds and domestic individuals which have wandered from the farms" ..... hunting turkeys and small game confined to Thursdays, Fridays and Saturdays.
- **1929** – Game Commissioner Ross Leffler announces plans to establish the world's first turkey propagation farm in central Pennsylvania ..... commission takes initial steps to close certain counties to turkey hunting ..... bows and arrows legalized.
- **1930** – Commission purchases 938 acres of contiguous farm and forest land in Juniata County for wild turkey farm ..... later purchases increase overall tract to 1,121 acres with more than 500 acres enclosed by a 9-foot fence ..... hunters credit the PGC for aiding the wild turkey emergence via the artificial breeding program but the truth eventually revealed that it was little but a put-and-take resource.
- **1931** – Maximum price for gamelands purchases set at $10 per acre.
- **Mid 1930s** – Agency begins land acquisitions beneficial to turkeys ..... planting grains and legumes, mainly in southern counties, bolsters food sources.
- **1936** – Wild turkey mating areas are established to improve quality and wildness of turkey farm stock ..... gobblers flew into 10-acre fenced area to mate with wing-clipped hens ..... first year's production of 4,431 eggs yields 1,428 hens and gobblers ..... gamelands system grows to 500,000 acres on 100 tracts in 52 counties.
- **1937** – Turkey calls legalized.
- **Early 1940s** – More meaningful wildlife research begun largely with federal Pittman-Robertson funding garnered from sportsmen's self-imposed excise taxes on sporting arms and ammunition.

• **1942** – 21 similar mating areas established across the state.

• **1945** – Juniata Turkey Farm closed because of lack of space to grow, contamination, inadequate facilities and poor soil ..... operation moved to Lycoming County.

• **Mid to late1940s** – Return and maturation of logged forests continues ..... turkeys begin to show up in areas never stocked ..... expansion incorrectly attributed to stocking, not improved habitat.

In 1945 the Loyalsock Turkey Farm was established in Lycoming County.

• **1949-1950** – PGC Biennial Report states: "Turkeys are increasing noticeably throughout their entire range due to more liberal and extensive stockings. The strain is as wild as can be maintained and even farm-reared birds are almost unapproachable."

Twenty-week old turkey farm stock awaits release into Blair County in October 1952.

• **1951-1952** – Biennial Report announces plans to "determine why wild turkeys have declined on the long-established range of southcentral Pennsylvania, to determine the value of farm-reared birds and investigate using artificial insemination at the Lycoming County turkey farm.

• **Mid 1950s** – Biologists estimate turkey range grew from about 2 million acres to 13 million acres ..... commission biologist Harvey Roberts states: "A sobering factor in this rosy picture is that the bulk of the turkeys are now being killed on the newly-extended range in the northern half of the state (and) the

Pioneer hunter Billy Neal, then 82, holds a turkey he shot with a double-barrel muzzleloading shotgun in October 1954.

southcentral portion, which only a few years ago represented (Pennsylvania's) entire turkey range but now produces comparatively few birds."

• **1954** – First statewide turkey season in decades ….. 10 or so counties still with no natural-occurring turkeys.

• **Late 1950s** – Trap-and-transfer projects stimulate population expansions on a county-by-county basis.

• **1960-1970** – 650 turkeys trapped in northcentral counties and released elsewhere.

• **1962-1963** – Biologist Arnie Hayden conducts research showing that turkeys could withstand lengthy periods without much food and not suffer reproductive setbacks ….. fellow biologists agree, concluding that no benefit is served by large-scale, winter feeding programs.

• **1968** – Pennsylvania's initial spring gobbler season held from May 6-11 …. estimated previous fall's population of 60,000 birds and overwintering numbers at

Turkey hunters set up to call in this 1950s era photo.

Turkeys took hold in good numbers in many regions in the 1960s.

30,000….. one-third of population gobblers ….. estimated kill of 1,636 toms.

• **1969** – First hunter education course required for youth under age 16.

• **1970s** – 900 birds trapped and relocated ….. unoccupied ranges include the state's northeast along with the southcentral counties and the Alleghenies.

• **1972** – Spring gobbler season expanded to two weeks ….. hen or gobbler permitted in fall (first time since 1915) ….. daily closing time for spring gobbler hunt moved from 10 a.m. to 11 a.m.

• **1973** – Compound bows legalized.

• **1979** – PGC commits more resources to accelerate trap-and-transfer program.

• **1980** – Wildlife managers from other states begin their own trap-and-transfer projects upon learning of Pennsylvania's

Game Commission personnel prepare to release trap-and-transfer turkeys from the northcentral region into Pike County.

success ….. after 50 years, turkey farm concept abandoned after production of more than 200,000 gobblers and hens ….. Lycoming County game farm switches to ring-necked pheasants.

• **1984** – Spring gobbler season extended through four weeks.

- **1985** – Nine Turkey Management Areas established.
- **1987** – Trap-and-transfer work concluded with exception of several localized efforts from the mid-1990s through the winter of 2000-2001.
- **1992** – Fall turkey hunters required to wear 250 square inches of safety orange while moving ….. unlawful to hunt turkeys with shot size larger than No. 4 lead or No. 2 steel.
- **1993** – Spring turkey hunters required to wear 100 square inches of safety orange on their heads while moving.
- **1996** – Record 241,613 hunters take to the spring woods ….. record 250,377 hunters hit the fall woods.
- **2000** – State's turkey population estimated at 400,000 birds.
- **2001** – Record spring harvest of 49,186 longbeards and jakes ….. blinds made of manmade materials legalized.
- **2003** – Spring hunter success rate 17 percent ….. turkey hunting blinds legalized.
- **2003** – 11 gun-related incidents involving spring and fall turkey hunters ….. no fatalities.
- **2004** – Calculated spring harvest of 41,000 toms ….. first Youth Wild Turkey Hunt held April 24 ….. second spring gobbler license ($20) approved ….. spring harvest estimated at 41,000 birds ….. thought being given to extending spring turkey hunt to all-day affair ….. turkey hunters continue to seek changes in mandatory blaze orange regulations ….. Sunday hunting controversy resurfaces ….. innovative youth hunting programs presented.

Season 2001 brought a record spring harvest of 49,186 gobblers, such as the Wayne County bird taken by Bob Tucker, Jr. of Coopersburg.

- **2005 and Beyond** – Only time will tell ….. but with state and national turkey federations at the helm, smooth sailing should prevail.

CHAPTER 2

LOOKING BACK

The Turkey's Long Road to Return

The turkey hunting world lost a good friend on August 6, 2001. Renowned biologist Arnie Hayden passed away at age 68, leaving behind a legacy of contributions to the return of the wild turkey across Penn's Woods and beyond.

Several months earlier, following a morning gobbler hunt in Tioga County, I stopped by Arnie's Wellsboro home for lunch and what would be my final visit with the amiable gent.

Biologist Arnie Hayden seeks response from radio-tagged turkeys in Tioga County.

We talked, of course, about turkeys and his remembrances of serving with the Pennsylvania Game Commission from 1961 until his retirement in 1995.

I sat back with a glass of iced tea, flicked on my tape recorder, and let him talk about his favorite subject. The interview proved to be invaluable.

"Finding turkeys wasn't all that easy back in '61 when I hunted them for the first time," Arnie began. "I went to Cameron County and had to hunt some pretty tough country because the thought then was that the only way you'll even hear a gobbler is if you get into some remote section."

He recounted hiking into a sector of contiguous forest covering seven miles of hollows leading to another pair of 4-mile hollows far from the truck. Few people ever penetrated the area so nobody really knew where the birds hid out, but the reputations of the birds as hermits were believed by most hunters. The big woods country held some hens and gobblers but none were found north of Mansfield in the late 1950s and into the 1960s. So, too were the southwestern and southeastern counties devoid of turkeys, save for individuals and hunt clubs who raised and stocked them as put-and-take fare.

"That area in the (southcentral counties)," Hayden recalled, "had only hunting camps used in deer season and turkeys were there. Today that whole area is full of houses and turkey numbers have declined dramatically."

Few Hotspots

Hopscotching the state, Arnie listed Bedford through Franklin counties in the southcentral region as a former hotspot along with northern counties from the Allegheny River in Warren eastward. The western rim of the state from Washington to Erie counties had no turkeys bellowing from their roosts nor did the Poconos and the rest of the Northeast. However, trap-and-transfer stock was placed in Carbon County and eventually spread into northern Schuylkill. From 1965 to 1970, however, Wayne and Susquehanna counties received trap-and-transfer birds and flocks were bolstered by "migrants" from New York. The numerous farm counties, including Bradford and stretching into the southeast, were avoided when it came to trap-and-transfer methods. Even biologists believed stocking birds in areas without vast contiguous woodlands would fail.

A New Game

In 1968, when the initial spring turkey hunts were approved by the board of commissioners, biologists wanted to close fall hunting to help build turkey numbers. But commissioners and other non-biologist Game Commission leaders ignored the request because license-buying hunters insisted that they wanted fall hunting.

"We had a tremendous turnout that first day," Hayden recalled with a smile. "I

Pat Mallory, then 13, shows fellow hunters the longbeard he shot in Huntingdon County in 1970.

counted 97 cars and trucks on Armenia Mountain (Potter County) but within a few years it was down to about 25 cars on the opener."

"The good news was that hunters came from all over the place," said Hayden. "The bad news was they didn't do all that well because they weren't sure how to hunt them in spring."

But for some hunters who were sufficiently woods-wise and could handle a box

call, (the only call available at the time, Hayden noted), "the birds were really gullible because they'd never been called to before." Yet, spring hunter numbers dwindled as participants soon learned that these birds weren't pushovers. The basics that every hunter knows today were mysteries back then.

Biologists requested closing the fall season in 1968 – first year of a spring gobbler hunt – but commissioners and many hunters vetoed the proposal.

The Game Farm Mindset

The most important and most controversial action by the commission was the hard-fought battle to kill the game farm turkey mindset of most hunters and commissioners. Add to that the strong belief that privately raising turkeys and setting them free was the way to go. Dumb birds were easier to hunt and, as was later learned, they stimulated in-breeding and carried diseases. But most of all it was the stupidity of the farm-raised birds that finally changed the minds of doubters.

Hayden recalled: "I remember one morning back in 1985 I was driving on a back road near Wellsboro and there were turkeys in the road. I stopped and grabbed my camera, figuring I could sneak a few pictures but when I got out of the car they stood there and looked at me. I walked right up to them, then I grabbed a club I had in the car and clubbed a couple to death. They should never have been released in the first place."

"Turkey harvest figures in the past were somewhat sketchy," Arnie admitted. "So we did calculated harvests like we do for deer today."

Early on, only about 18 percent of all (successful) hunters sent in report cards, largely because they used the same cards to report deer and bear kills and didn't have a card left to report turkey kills. By 1972 that figure jumped to a 50 percent reporting rate for the spring hunt and 40 percent in fall.

A New Approach

Over the years the game farms finally closed and trap-and-transfer operations were set in place. In varied counties fall hunting was shut down to allow the flocks

The Loyalsock Turkey Farm was finally closed in 1980.

to populate and extend their ranges. But Arnie and fellow biologist Jerry Wunz held the strong belief that the creation of nine Turkey Management Areas was the changing point in keeping tabs of turkeys and getting flocks established.

The zones enabled a more ecological method of management, including the controversial banning of fall hunting in some zones, when necessary. Now, however, instead of confining management efforts to counties alone, larger regions have been put in place and closures could be applied to specific zones. The concept exists through today in the short-term closures in certain Turkey Management Areas (now called Wildlife Management Units) where turkey numbers still need protection from overharvesting. Shortening or lengthening the fall hunt periods in varied WMUs continues as an important management tool.

Net Results

Yet another benefit to the turkey restoration program was the advent of the cannon net, which replaced the antiquated wire "box traps" used in the early part of the trap-and-transfer project. These traps, Hayden explained, measured 35-40 feet in length and were covered in cotton camo. Turkeys went in to eat the bait but couldn't get back out. The cannon netting procedure – used on ducks in the Midwest – speeded up the process and snagged many more turkeys than the outdated box trap.

An error in judgment, however, created a glitch in the operation; the birds were initially caught and released in summer. In many counties predators and hunters took the young birds in the fall season. In Perry County, for instance, "turkeys were annihilated," said Hayden. In 1964 winter trapping was instituted.

By the late '80s turkeys were showing up in good numbers in suburban regions, such as these Bucks County birds checking out bird feeders on the edge of a woodlot.

Asked the most dramatic and meaningful lesson he learned about turkeys over the years, Arnie quickly answered: "It is the ability of turkeys to adapt to many different situations."

"Turkeys in farm country and even suburbia was once considered impossible. They lived in woods, not fields. But we learned differently over time."

One of his bigger frustrations was in dealing with Michigan in a trade agreement involving 200 wild hens and gobblers in 1979-1980 in exchange for snowshoe hares. But the Game Commission never received the varying hares. That was followed by a trade with Michigan for Sichuan pheasants, a Chinese variety which thrives in brush country and was to be used for breeding stock in exchange for another 200 Pennsylvania turkeys. But the deal went south once more. Michigan sent all male pheasants which had little use but were cross-bred with ringneck hens. The project was a total failure and Hayden didn't hesitate to brand the Michigan game department as "dishonest and unethical."

Arnie concluded the interview with a surprising observation: "With all the problems we had, if you'd have asked me what the future of turkeys was back in 1961 I'd have said "They probably won't be here in 20 years."

That was more than 40 years ago. Today as many as 375,000 to 400,000 turkeys leave tracks across Penn's Woods and beyond, thanks largely to Arnie Hayden and other dedicated biologists who never gave up the fight to restore these fascinating gamebirds to their original stomping grounds.

CHAPTER 3

TRAP-AND-TRANSFER

The Wild Turkey Success Story

The Pennsylvania Game Commission's initial trap-and-transfer programs began long before methods were developed to make the venture time and cost efficient. The first traps relied on getting the wild birds inside a large "walk-in" container by following a bait trail. Trouble was, only a smattering of birds was caught; not enough to begin any viable restoration program.

"Wild turkey trap-and-transfer got its start in Pennsylvania in the 1920s with that kind of trap," said Mary Jo Casalena of Bedford, the Pennsylvania Game Commission's current turkey biologist. "Wildlife trap-and-transfer projects have been used throughout North America by government wildlife agencies and the National Wild Turkey Federation but it really came into its own in the 1960s and '70s when more than 1,500 turkeys were trapped in Northcentral counties and released throughout the state."

A rocket net is launched across a flock of turkeys as step one in the trap-and-transfer program.

Most, but not all, turkeys were captured on grounds closed to hunting, including some private lands where landowners gave permission and on properties enrolled in cooperative public access programs. All transferred turkeys were leg-banded immediately upon capture to facilitate dispersal records and spring harvest rates.

Rocketing to Success

More efficient was the PGC's efforts to capture turkeys using a rocket net that, when detonated, shoots a large nylon, mesh over the birds. To entice the hens and

The Pennsylvania Chapter of the National Wild Turkey Federation continues to supply transfer boxes for the trap-and-transfer operation.

gobblers into a suitable field situation for netting, only bait is first set out. Once a flock begins using a cracked-corn-baited site regularly, a trapping crew sets up the rocket net. A trapper then waits in a camo blind for enough birds to come in. When they settle onto the bait, the net is detonated. Three small rockets shoot the net over the turkeys.

Crews attempt to quickly remove the birds, band them and put them into turkey transport boxes to ensure safe passage to a release site. The boxes provided the birds with sufficient room to move while preventing them from flapping their wings and subsequent injury. Areas receiving birds were typically closed to fall hunting until viable populations were established.

Help from Hunters

The cost of the trap-and-transfer project was not borne by the PGC alone.

"The National Wild Turkey Federation is proud to be a part of this," said Rob Keck, a former Pennsylvanian now the CEO of the NWTF. "Since 1985, our volunteer members have contributed more than $2 million toward wild turkey management in Pennsylvania and we're quite proud of that."

PGC biologist Bill Drake of Kane, who worked on the turkey program for 10 years before his retirement in 1999 following 29 years with the agency, said the NWTF "was an instrumental force in moving the trap-and-transfer program along."

"They provided volunteer and financial support but they also took the lead in influencing other hunters when it came to resistance to closing fall seasons for a time," Drake said. "The Pennsylvania chapter (of the NWTF) was a big part of the management team and was always conservative in backing positions that prevented overharvesting. I liked that."

"But politics was around back then, too," Drake chuckled.

"Arnie (Hayden) and Jerry (Wunz), especially Jerry, sometimes took unpopular stands that game commissioners just didn't agree with. Jerry fought the battle to get rid of game farm birds and that didn't set well with some commissioners. It got him into hot water more than once but he stood by his guns as to what was the right thing to do – and he won."

Indeed, we all won as the success of trap-and-transfer urged the PGC along the road to wise management of the Eastern wild turkey as proven by today's viable populations in nearly all counties across the Keystone State.

Trap-and-Transfer Continues

Although the Pennsylvania Game Commission closed shop on most of its trap-and-transfer operations in 1997, this winning technique for restoring hen and gobbler populations is not yet extinct. In portions of Wildlife Management Units 5A, 5B and 5C north and west of Philadelphia, trapped birds have been relocated. The restocking area includes portions of York, Lancaster, Berks, Chester and Bucks counties.

"This project began in 1984 and continued intermittently until 1997," said project coordinator Mary Jo Casalena. "But in this region populations didn't

Relocated wild turkeys quickly took hold, including farm and woodlot country where transfers were not expected to do well.

expand to other suitable habitats as we expected."

With a basis for concern, in 2000 Casalena directed research to identify sites in a 20 square mile region of the Southeast offering forest, urban-commercial land and agriculture property. Eventually 15-20 turkeys -- with sex ratios of three hens per gobbler or two jakes per adult male -- were trapped and released. Turkeys were chosen from at least two trapping sources for genetic variation in the suburban woodlots and farmlands where they were eventually set free. Birds were also released in habitats with existing populations to encourage interaction, again for the sake of genetic variation. By project's end in 2003, 515 birds (169 males and 346 females) were translocated onto 21 sites.

Casalena explained that "although small flocks of turkeys existed in a few forested areas of southeastern Pennsylvania, the areas are fragmented, making it difficult for the birds to move or disperse into other suitable habitats."

In some counties turkeys didn't expanded their ranges as anticipated due again to fragmentation of habitat (rivers, suburbia and major highways).

Although restocking efforts have been canceled for now, conservation officers and biologists will be keeping tab of range expansion, turkey numbers and annual harvests in the region.

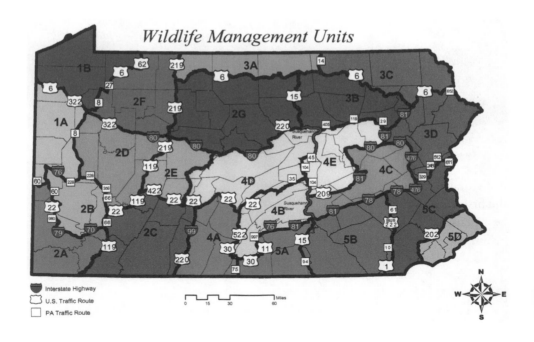

CHAPTER 4

THE FIRST SPRING TURKEY SEASON
Satisfaction the First Time Around

Harvey Fouse and his father Fred both shot longbeards in Huntingdon County on the opener of the first gobbler season in 1968.

Naturally, the usual and predictable complaints from skeptics were heard when the first spring gobbler season in 1968 was first proposed. But it didn't take long to hear praise from those who actually tried spring hunting.

Surprisingly, hunters who didn't bag a bird or even see one said they were satisfied simply to hear a gobbler sound off at daybreak. This vocal display by the tom turkey quickly became the key to the hunt's instant appeal. Just hearing a gobble or, better yet, have a bird answer your call was a thrill. Hunters, particularly those disheartened after going through the previous fall season without as much as a glimpse of their elusive quarry, were amazed at the abundance of gobbling birds that populated our prime turkey range in the spring.

Credit the Pennsylvania Game Commission for doing its part to ensure success by exhaustive research before spring turkey hunting was recommended as feasible. The experiences of other states that already had spring hunts enabled the PGC to make accurate predictions on the probable outcome of our initial season.

To make sure all was on the right track, commission studies included surveys to determine hunting pressure, hunting success, hunter behavior and effects upon the turkey population. The first of these was a count of hunters conducted over established routes throughout the turkey range by Game Commission and Allegheny National Forest biologists. The findings showed spring hunting pressure was less than half of that occurring during any comparable day of the fall turkey season. Pressure was lowest in the vast northcentral turkey range and greatest in ranges closer to heavily human populated regions of the state.

As was suspected, 50 percent more hunters were counted on the last Saturday of the first 6-day season than on the opening Monday. The season was purposely opened on a Monday to avoid the possibility of excessive pressure that might have occurred on a Saturday debut.

The second part of the survey was completed by hunters who found post cards left on their vehicles' windshields or were given one directly by a game protector. The cards revealed the average spring hunter spent two mornings afield, heard 3.3 gobblers and saw 1.4 turkeys. Twelve percent of the respondents reported bagging a turkey. Hunter success was greatest in northcentral Pennsylvania and lowest in the Southeast.

The third survey involved findings from game protectors across the state. They noted that hunter behavior was generally good with incidences of only six illegally killed hens. They were voluntarily turned in by hunters for payment of one-fourth the listed penalty.

Other findings included: Only 36 turkey nests were reported encountered by hunters; game protectors examined 275 gobblers and two legal bearded hens; field counts factored with biologists' calculations revealed that 1,636 turkeys were harvested in the first spring season.

Yet another survey was the annual winter census. This involved locating and counting the flocks by their tracks in the snow while they were concentrated on their winter range. The question of whether the new season had harmed the turkey population was answered: It didn't. If it had, regardless of the great challenge provided, future spring hunts could not have been justified.

So successful was the initial hunt in 1968 that the following year's season was lengthened to seven mornings with a Saturday opener. Hunters reported toms pushing 20 pounds in the spring with several reports of longbeards in the 25-pound category. This was a rather pleasant surprise to many hunters who expected turkeys to be somewhat scrawny after a long winter but instead learned that the birds were plump and good eating.

The first spring gobbler season's debut in 1968 continues as one of the Keystone State's top success stories.

A winter survey following the 1968 gobbler hunt showed that turkey numbers were not negatively affected by the spring season.

Author's Note: *The preceding article is an edited version of a report by Jerry Wunz, who was instrumental in leading the PGC's turkey restoration and habitat program along with Arnie Hayden. He served with the Pennsylvania Game Commission for 33 years, retiring in 1990. Jerry passed away on February 11, 2004 at age 78.*

CHAPTER 5

PERILS OF PEN-RAISED TURKEYS

"Polluting" the Bird's Wilderness

It took more than four decades for wildlife managers and hunters to learn an important lesson about wild turkeys: You can't raise them like farm-grown butterballs.

Failures in the raise-and-release projects in Pennsylvania and elsewhere demonstrated the folly of attempting to restore pen-raised turkeys to once-occupied habitats. The plan did little but make easy targets for owls, hawks, foxes and coyotes, no matter if the turkeys were set "free" as poults or adults. Yet, the practice still occurs today.

Pen-reared turkeys make easy targets for predators and may infect wild birds with diseases.

In places where turkeys aren't abundant, well-meaning sportsmen's organizations and individuals continue to attempt to increase turkey numbers by stocking them like pheasants. Birds bought from some game breeders aren't true eastern-subspecies wild turkeys. They're the products of selectively breeding wild and domestic turkeys to look like eastern wild turkeys. Genetically and behavior-wise, they're far less able to survive the rigors true-strain birds must endure.

"Captive rearing of wild turkeys eliminates the important learned behaviors that wild turkey poults acquire from their mothers," said Bob Eriksen, a National Wild Turkey Federation regional biologist for New Jersey and Pennsylvania. "Therefore, they're more susceptible to predation and less likely to breed successfully."

The Disease Factor

Indeed, most privately stocked birds are little more than dark, barnyard turkeys and are more likely to exhibit tame or even aggressive behavior toward humans than their wild counterparts. Furthermore, Eriksen and other biologists know that pen-reared turkeys can have negative impacts when they come in contact with wild stock. For one, they pick up diseases and parasites when raised in proximity with other gamebirds or domestic fowl. Exposure to diseases or parasites on the farm is but one example.

The pen-reared turkeys may at first appear hardy and may not become ill thanks to treatment with medication in their food or water. Upon release, however, the health of these turkeys suffers as they try to adapt to the wild without medication. Diseases are then passed on to the wild birds with which they come in contact. If the birds survive and reproduce, their inferior genetic qualities then mix with those of truly wild turkeys and the entire genetic pool suffers. Consider it a "pollution" of their wildness and hardiness.

"The Pennsylvania Game Commission allows people to possess pen-reared "wild turkeys" but does not allow release of the birds without a permit," Eriksen notes. "(Doing that) without a permit can result in fines."

Even captive turkeys are colorful and fascinating to watch, Eriksen admits. Turkey hunters often keep penned turkeys to study their vocabulary and practice calling techniques – or just for the fun of watching them.

Enjoy them, says Eriksen, but never, never release them in the wild.

EARL MICKEL
Beach Lake, PA

Earl Mickel is the author of *"Turkey Callmakers: Past and Present."* He hunts in Wayne and Pike counties and each year pursues gobblers in a dozen other states.

• For more information on Mickel's insightful and historical turkey callmaker books, contact him at mikspiks@ptd.net

Early Call Makers Set the Stage for Today's Industry

The first call makers were Indians who used a green leaf or hollow reed, later progressing to a single turkey wingbone or embellished to two or three wingbones fastened together.

The first advertised turkey call was a B.G.I. yelper named for the British Gun and Implement Company who sold them in the late 1880's. Today's value is about $600. It's a fairly decent suction yelper when compared with current models.

Henry Gibson patented the first box call in 1897. Gibson's box calls are worth about $2,000 currently. They leave a lot to be desired as a turkey call compared with today's quality box calls, but back then they were effective.Pennsylvania's best known callmaker of his time – D.D. Adams of Thompsontown – built and patented the double slate and peg call. By doing so he revolutionized the call making industry. His contribution was one of the great advances made to turkey call making. Many other Pennsylvanians have received patents and/or contributed largely to the industry but none to the degree of D.D. Adams.

D.D. loved the outdoors and turkeys were his bag. He got his first one in 1921 and hunted until his death in 1988 at age 81.

No one is sure when D.D. made his first call, but they were tube-type calls. He initiated slate calls in 1974 or 1975. These were simple round slates with a wood holder and usually round or octagonal bases. They came with a very ornate striker, usually checkered, that D.D. turned on his lathe.

He started using flowerpot bases in 1980, about the same time he began making the now famous double slate. The bases were bought locally and varied according to what was available. Some have amazonite bottoms and others can be struck from either side. He was an experimenter. About the time D.D. introduced his double slate he also began making a longer and more streamlined wood striker. He also made Plexiglas and wing bone strikers.

This crude wooden friction call and slate striker was one of the earliest made. (Maker not known.)

The Biology of the Bird

A mother hen keeps watch over her chicks only days from the nest, captured on film by WCO Tim Flanigan of Bedford.

CHAPTER 6

TURKEY BEHAVIOR AND BIOLOGY: 101

The Eastern Turkey's Life and Lifestyle

My friend David Hale of the famed Knight & Hale Game Calls company once remarked that "a true Grand Slam should be shooting four Pennsylvania gobblers in a single season."

I thought about that a number of times when hunting in other parts of the nation holding the Osceola, Merriam's, Rio Grande and Eastern turkeys. Not that these birds – or the lesser-known Gould's subspecies of the southwest – are pushovers; rather I've found them to simply offer a different type of hunting in habitats as varied as Florida swamps, Texas scrub and the mountains and plains of the American West and Midwest.

Eastern gobblers are the most widespread of the four subspecies, thanks largely to trap-and-transfer cooperation between state game agencies over the years. They're also arguably the most pressured turkeys. Hence, they hold the reputation as being more difficult to bag and tag, especially Pennsylvania populations.

Hally's Callers Joe Hall, Jr. and Joe, Sr. of Doylestown travel the outdoor sports show circuit with their unique display of five subspecies of turkeys, their habitats and audio of their calls.

It's estimated that as many as three million Eastern turkeys inhabit all or portions of 37 eastern, western, midwestern and southern states. They dwell everywhere from wooded mountain ridges to lowland swamps and have adapted to farmland and even suburban living.

Pennsylvania is home to only the Eastern bird – some 350,000-400,000 of them. An Eastern Grand Slam (four birds in a year) for a Pennsylvanian hunting his or her homeland isn't possible. It would require a visit to another state holding Eastern turkeys. As this is being written, three Pennsylvania toms are legal in a year's time; one in spring with the regular license, another spring bird (with a special permit) and one in fall.

Hale is probably correct in his assessment that Pennsylvania turkeys are among the toughest to hunt. Grand Slams aside, just taking one Eastern gobbler every two or three years is a challenging venture. I know hunters who have gone afield for 20 or more years and still haven't scored on a spring longbeard or jake.

The More I Know the Less I Know

Each time we step out in search of a gobbler they teach us new tricks and, as a friend once proffered after his first longbeard in 14 years of hunting, "The more I learn about them, the less I know. I've quit trying to figure them out."

That somewhat convoluted logic has a ring of truth. Yet, knowing the basics of biology and behavior of the Eastern turkey is, like deer hunting, a necessity for making all of us better hunters. Today, most lessons in what to do and not to do come from The Outdoor Channel, videos, books and magazine articles. Although valuable, the most profound lessons are learned by spending enough time in the woods to learn as much as we can about this magnificent bird.

With that firmly in mind, welcome to Turkey Behavior and Biology: 101.

KNOW YOUR WILD TURKEY TOM

Considered among the grandest of game birds, the American wild turkey has many characteristics that distinguish it from other fowl. The unmistakable snood, caruncles, head coloring, and beard truly set it apart.

TAIL FEATHERS OR RECTRICES
There are usually 18 present, but a gobbler can lose a few when fighting. Tan- to brown-tipped on Eastern, Rio Grande and Osceola subspecies; ivory-tipped on Merriam's.

TAIL COVERTS
Tip colors vary with subspecies.

BACK AND BODY FEATHERS
Provide insulation and shed water. When upraised they refract sunlight to add to a strutting tom's grandeur.

PRIMARY WING FEATHERS
Marked by distinctive white bars (less barring on the Osceola). Toms rub off wing tips with extended strutting.

SPUR
Most spurs are black, some have red or blond tints. They appear as a short button on a jake; just less than a one-inch straight spur on a two-year-old; and as a sharp, curved, one-inch or longer spur on a three-year-old tom.

EAR OPENING
No flap to funnel sounds, but a tom hears extremely well.

HEAD CROWN
Predominantly white during the spring, sometimes with a reddish tint.

EYE
Set into the side of the head for monocular vision; a slight turn of the head allows a 360 degree field of vision.

SNOOD
Long and prominent on a mature gobbler, but no known function.

MAJOR CARUNCLES
Large and fleshy. Engorged with blood during the spring.

BREAST FEATHERS
Black tips give a gobbler a coal-black appearance.

BEARD
Three to four inches on a jake; seven to nine inches on a two-year-old; and 10 inches or longer on a three-year-old gobbler. Thickness varies. Some toms have multiple beards.

FOOT
Three long toes. The middle toe measures 2½ to 3½ inches on a gobbler.

Know the Difference Between Hens & Gobblers
It's easy to distinguish a gobbler from a hen by differences in their size, color, heads, and other characteristics.

Those Spectacular Feathers

As to identification, the wild turkey needs little introduction, although comparing a farm-raised butterball to a wild and wary tom is a lesson in dissimilarity. Barnyard birds are heavier, have bigger heads, shorter necks and stumpy feet. And they're dumb. Real dumb. That's why calling someone a "turkey" is an insult but to a dyed-in-the-camo turkey hunter, it may well be a compliment.

Adult gobblers are slimmer and stand about three feet tall with their mates reaching about two feet, give or take an inch or two. Hens of 9-12 pounds are the norm with adult toms carrying 18-22 pounds, along with the occasional 23-25 pound longbeard.

Plumage on longbeards – composed of 5,000-6,000 varied feathers –is largely black and brown but under cloudy-bright sunlight a spectacular array of colors appear. Check the cover photo of this book as example. The iridescent tints of blue, green, red, copper and mahogany are, indeed, stunning.

A hen's feathering is notably duller but she also shows a bit of iridescence when viewed in the proper light. The head of a hen is a blue-gray although I've had excitable boss hens answer my calls and noted a deeper blue color on their bristly heads.

To some people turkeys are simply black but a close-up view shows the spectacular iridescence in their plumage.

Pure albinism in Eastern turkeys is a rarity. More common is a smoky, gray mutant – sort of an incomplete albino – with dark eyes and feather tips. Both hens and gobblers may show the aberrant color.

Colorful Caruncles

Anyone who has ever had the privilege of watching a strutting gobbler come to a decoy knows the Kodachrome display well. A tom's head during its time of lusty gobbling and calling is either white or red – and sometimes royal blue. Mostly, however, it will be a mix of white with red and blue patches.

A close look also reveals the strange-looking caruncles which Webster describes as "an outgrowth of

The gobbler's snood and caruncles provide it with ever-changing neck and head coloration.

flesh" on the head and neck of toms. A curious part of the caruncles is the long, thin snood originating between the eyes and draping over and down the beak. When a jake or longbeard becomes excited the caruncles and snood fill with blood and the latter becomes firm. Its purpose is believed to be purely visual and sexual.

Look for the Beard

Of course, everyone who's ever seen a wild turkey – or a domestic tom – is aware of its beard. Those who are not turkey hunters or are otherwise disinterested in wildlife may presume that a "beard" of any sort would grow directly below the chin. A gobbler's beard, however, hangs from the upper breast.

Surprising to some is the fact that a turkey's beard is composed of feathers, not hair. Only mammals grow hair. Beards are actually modified feathers which continue to grow throughout a gobbler's lifetime and may reach 12 inches, although most adults are 8-10 inches.

A yearling jake's beard will grow to 3-4 inches or so the first year, with some variation, although on some jakes the beards are barely visible as they're buried in the large chest and breast feathers. Not until six or seven months after hatching is a mere hint of a beard seen. This requires a close look when a jake comes to a call as even some year-old toms lack much of a visible beard.

If a beard's not visible, don't shoot. Pennsylvania law permits the killing of "bearded birds only," not only male birds. Strange that a bearded hen is considered legal but a non-bearded jake is not.

Gobblers' beards grow 3-5 inches annually but will become stunted when the end bristles are broken as the bird wanders and feeds. Freezing snow accumulating on the

Some gobblers have two, three or even more beards.

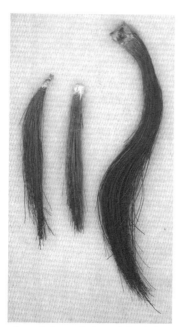

An adult tom's beard (right) as compared to a jake's beard (center) and a hen's beard.

beard will also result in an inch or two of beard loss over a harsh winter. Any beard of 10 inches or larger is notable. One Eastern turkey shot in Alabama in 1985 held a 16.9 inch beard.

The density of a beard is determined by the number of filaments making it up. These unique beards are sometimes referred to as "brooms" or "paint brushes." Most beards hanging from older birds are entirely black but some will show blonde to reddish-brown, especially younger toms.

Although far from rare, it's a bonus when a gobbler is taken with two, three or more beards. One Wisconsin Eastern subspecies tom had eight beards totaling 71 inches.

Hens also grow beards from time to time and the phenomenon occurs more than we might imagine. It was discovered that as many as eight percent of hens were bearded in a New Jersey study and up to 29 percent has been recorded in other states.

It's easy telling a small hen from a jake via its size and coloration. Although bearded hens are legal here, it's this writer's belief that ethical hunters will pass on these curiosities. If a hen tending eggs is shot when it leaves the nest to feed, its eggs will surely not hatch.

Nubs and Spurs

The first thing a successful hunter does, once the bird has stopped flopping about, is to quickly inspect the beard, spread the tailfeathers and rub his or her fingers across the spurs – sharp, bone-like projections growing on the back of the lower legs. A jake has little more than roundish nubs at the outset but with each year the spurs become longer, eventually developing into curved, sharp weapons which are used in fights.

More than one hunter has grabbed onto a flopping gobbler's leg only to drop it like the proverbial hot potato as a sharp spur opens a gaping would in his palm or fingers. I know several excitable hunters who have taken trips to hospital emergency rooms after a successful gobbler hunt.

Spurs this size are an exception, growing only on old longbeards.

Two inches is about the maximum spur growth although it must be noted that such a spurred bird is a once-in-a-lifetime trophy, a true "limb hanger." The oft-used term denotes that the bird can be hung from a horizontal limb by only its curved spurs. Inch-plus spurs are more the norm as on older birds they will wear down a bit in rocky habitat. Most spurs on adult toms are a brownish black but some birds possess reddish or off-white spurs. Only rarely will a hen grow tiny spurs.

Preseason scouting trips should focus on roost sites and travel routes at fly-down time.

What's for Lunch?

Turkeys travel on their stomachs, especially in fall and winter when food sources dwindle. Spring brings the renewal of varied foods and even though hens remain as hungry as ever, toms will merely pick here and there. Like a rutting buck, a longbeard will happily sacrifice supper for sex.

Scouting trips prior to the spring hunt should focus on turkey sign such as scratchings, dusting sites and roost trees, not necessarily food sources. Most of the previous autumn's soft and hard mast has been devoured. Once the season gets into swing and greenery colors the woods, however, tender green shoots, buds, tubers and nuts that haven't rotted compose the menu. Scratchings in the forest leaves turn up a variety of morsels from spiders and millipedes to slugs and the occasional worm or salamander. If it's edible and can be swallowed, a turkey will probably eat it. Gourmets they are not.

When fields begin to green up turkeys increase their use of fields for insects, such as grasshoppers.

Most hunters have watched how turkeys feed, scratching with both feet, then stepping back and pecking here and there on the bare ground. A troop of turkeys moving across a woodland floor will leave a trail of obvious scratchings.

Fall's crop of small, triangular beechnuts are irresistible to turkeys.

As fields green up, the welcome hordes of grasshoppers and crickets become available and of interest to the growing poults and the old toms and hens. Come late summer and fall, the pantry is once again filled with seeds, fruits, berries and nuts. Favorites include the soft mast such as barberry, pokeberry, crabapple, autumn olive, dogwood, wild cherry, wild grape, thornapples and seeds from sunflower, corn, oats, foxtail and more. By winter their diet changes to fall's leftovers, greenery growing in spring seeps and an occasional insect larva.

Despite the leggy cuisine they so enjoy, a turkey's year-around diet is 90 percent vegetal. Given one food in fall, the turkey's favorite dining spot will be beech woods where the tiny, triangular nuts are abundant. A flock will travel a long distance to find such a treat. Second, perhaps, are acorns which also are the favorites of deer. Deer, however, crunch acorns like my wife annoyingly crunches on hard, Pennsylvania Dutch pretzels. Turkeys, however, simply stretch their elastic necks upward and swallow. I wish my wife could do that.

The internal crop below a turkey's throat stores the food which then passes into the hard gizzard and efficiently carries out the chore of breaking them up for easy digestion. The next time you field dress a bird, run a knife through the gizzard and, if you're near a creek, rinse the organ. You'll be amazed at the numbers and sizes of stones and gravel in the

These stones, which aid digestion, were removed from the gizzards of two eastern Pennsylvania gobblers.

gizzard which help grind the food into digestible size. The stones are picked as the bird's feed in the leaves or are taken from gravel roads and trails.

The Eyes Have It

Time was when turkey hunting stories embellished the acuity of the wild turkey's eyesight as the equivalent of a 10x optical scope. Over the years, however, I've never come across any studies supporting the claim. Yet, there's not a turkey hunter around who doesn't know that movement is the number one giveaway to their presence, even at long distances, thanks to a turkey's phenomenal vision.

One blessing all turkeys possess is an ability to pick out detail in an object. Because its eyes are on the sides of the head, a turkey possesses monocular vision. That is, it can see 300 degrees without moving its head. With but a slight twist of the neck it is instantly able to see the remaining 60 degrees.

That puts a hunter walking through the woods or set up against a tree at a certain disadvantage. One false move and you may be out of business.

A turkey's night vision is poor. When blown out of trees by storms or spooked by deer hunters heading to their stands before first light, turkeys stand the risk of being injured by

Turkeys live by their eyesight but their hearing is also acute. Note the tiny ear opening behind and slightly above the eye.

crashing into limbs or misjudging the dark ground as they try to fly off into the unknown.

Like all birds, turkeys can see color. If they didn't, a gobbler's bright red wattles would serve little purpose at mating time. The eye also possesses a whitish nictitating membrane or inner eyelid which keeps dust, pollen and other irritants from the cornea and pupils.

Modern camouflage has played a huge role in helping hunters to "hide" from sharp-eyed toms and hens. It both breaks up the human outline and blends the hunter into his or her background. But make one foolish move in shifting your seat, scratching your nose or raising your gun and the jig is up.

Playing It By Ear

Compared to the ears of a deer or dog, it would not seem that the turkey's hearing ability could be very acute. After all, it lacks ear flaps to gather and concentrate sound waves and has but a straw-thick opening behind each eye.

Yet, as I've learned while hunting turkeys for more than 30 years and by experimenting with flocks that happen by my deer stand now and then, their hearing ability is acute. That alone is good reason to whisper or talk softly when hunting with a partner and to begin any calling series with rather subdued tree calls, yelps and cackles rather than using an introductory, high-volume beckon.

Field observations suggest that turkeys can hear low frequency sounds – such as another gobbler drumming – and more distant sounds than humans are able to detect. They're also able to pinpoint sounds, demonstrated by their ability to know exactly where a hunter's calls originate.

Smell and Taste

My friend Bob Clark likes to say: "If a turkey could smell like a deer, nobody would ever kill one."

How true.

As in nearly all birds (save for turkey vultures), taste and smell are not important senses for survival. In fact, turkeys might not want to taste some of the entrees they eat, such as slugs and millipedes. Yet, research hints that their low number of taste buds can probably detect bitter, acidic, salty and sweet foods, although not well.

The tiny olfactory lobe in front of the bird's brain interprets odors but it is an underdeveloped organ and is not believed to be very effective. In food selection, shape and color probably play a more important role than do taste or smell.

From Egg to Poult

Of course, no productive egg laying is possible without fertilization by a gobbler. Little known among hunters is that sperm is stored in the hen's oviduct. A single mating is usually

Hens begin laying in late April, depositing a single egg each day.

Incubation, a 28 day process, does not begin until the entire clutch is completed.

sufficient for fertilizing an entire clutch up to four weeks following mating.

The processes of egg laying and incubation are of special interest to spring hunters because it takes place prior to and during the season.

Hens begin laying in late April and continue to visit the nest site daily, depositing a single egg each day until a clutch of a dozen or so is ready for incubating. If she began incubation with the initial egg or two, they would hatch at varied times and create big problems in keeping tabs on everyone.

During the laying period a hen will scratch leaves over the nest when she goes to feed but, once she begins incubating, the eggs are usually left exposed for reasons that defy biologists. The normal incubation period lasts 28 days. When all the eggs have hatched (each chick takes a day or two to pick through the shell), mom and the kids quickly hit the road.

The hen will continue to brood them each night for about two weeks, at which time the youngsters are able to flit into low-branched trees.

Late spring into summer reveals the

After about two weeks, poults can flit onto low, vegetated limbs to spend the night.

general success of a hatching season as hens and their surviving chicks or poults often get together in large, nomadic flocks. People see them here and there as do predators.

It's not long after they hatch that the poults quickly learn the ways of the wild and instinctively freeze at the boss hen's warning call. Smart and obedient

One study showed that 53-76 percent of all poults perish before they're 2-1/2 weeks old due to bad weather and predation, the latter including the common great-horned owl.

birds await the all-clear signal before moving. Poults making the mistake of ignoring the calls end up as food for owls, hawks, foxes, snakes and other predators.

With abundant high-protein insect diets and seeds and other plant life, the poults grow to three pounds in their first three months. By seven months young hens weigh about eight pounds and junior gobblers grow to about 12 pounds, surpassing their mothers.

According to "The Wild Turkey – Biology and Management," published by Stackpole Books (Harrisburg; 1992), poult nesting success ranges from 31-45 percent, about normal for any ground-nesting species. Fifty-three to 76 percent of poults perish, mostly within two weeks of hatching. The life expectancy of a turkey surviving its first two weeks of life is still less than 1-1/2 years. Some turkeys have been known to survive more than 10 years in the wild. Annual turkey survival generally ranges from 54 to 62 percent.

In fall, the youngsters get their first tastes of nuts and berries

Most jakes have visible beards at 6-7 months of age and 3-4 inch beards by the following spring.

and continue to add fat and muscle to take them through the pending winter. Two dozen or more hens and young may travel together, scratching away on the woodland floor and gleaning anything edible in their paths.

By spring, jakes have grown short beards and most hens mate, then search out quiet areas in which to nest. Jakes will be fair game for the second time (the first during the fall hunt) and will travel together with a short-bearded crony or two, trying to figure out what the strange spring urge is all about. Hens will extend their ranges, often moving two or three times as far as adults and traveling beyond their former home rage.

Thereby they perform their intuitive duty of spreading into new territory and expanding Pennsylvania's turkey range.

Thanks to Pennsylvania Game Commission turkey biologists like Arnie Hayden, Jerry Wunz, Bill Drake and Mary Jo Casalena along with Bob Eriksen of the NWTF, the biology and behavior of wild turkeys – from egg to adult – is passed from scientist to hunter.

BOB ERIKSEN
Phillipsburg, NJ

Bob Eriksen is a NWTF regional turkey biologist for Pennsylvania
and New Jersey. He holds an M.S. degree in wildlife management
from Rutgers. His turkey hunts take him to Northampton and
Susquehanna counties.

WEATHER A MAJOR FACTOR IN YOUNG TURKEY SURVIVAL

There is surprisingly little variation in wild turkey egg-laying and incubation
dates between southern and northern Pennsylvania. Nest initiation by hens is
largely stimulated by photoperiod (the number of hours of daylight). While some
hens breed as early as mid-March, egg-laying does not occur immediately.

Most hens do not begin laying eggs until the middle of April. Even when trees
and ground cover leaf out early, the timing of wild turkey nesting does not vary
significantly. Studies conducted throughout Pennsylvania between 1953 and 1963
indicated that the average date of incubation initiation (when the hen begins to sit

on eggs after laying a clutch averaging 12 eggs) was April 28 statewide.

More recently (1999-2001), during a radio-telemetry study of wild turkey hens in southcentral Pennsylvania, the average incubation date for adult hens was found to be May 8 and juvenile hens began incubating on May 13. The opening day of the spring gobbler season is set to begin around the average peak of incubation, on the Saturday closest to May 1.

Hen turkeys are persistent in their nesting attempts. However, as is typical with ground-nesting birds, many first nesting attempts are lost to predators. When a hen loses her clutch she will often begin a second nest. Second nests usually contain fewer eggs than first nests and will hatch in late June or even July. These second nests are a great insurance policy against poor production in some years.

Most hens will re-nest if their first eggs are lost to predators.

Cool, wet weather in May can cause hens to abandon their nests when predators approach. Some biologists have theorized that wet hens give off more scent than dry ones, increasing the rate of nest predation in wet weather. So wet spring weather can influence the hatch by causing an increase in disturbance and nest predation.

After turkey poults hatch, losses are high in the first two weeks. Cool, wet weather in June can contribute to poult losses by reducing the ability of poults to thermo-regulate (keep warm). Even though the hen broods her poults often to keep them warm, wet weather dampens her feathers and makes it difficult for her to dry the young birds. Some poults die of exposure when this occurs.

Many poults are also lost to predators in the first two weeks. The young birds roost on the ground with their mother for the first couple weeks of life and losses can be high during that critical time. Rainy, wet weather in May and June can impact nest success and poult survival substantially.

Warm, dry springs are ideal for good hatches. Rain is much less of a potential problem when the air temperature is above 60 degrees.

No matter how you look at it, weather is the major factor in survival of young wild turkeys. When the chicks first hatch and throughout the summer, food is abundant.

If the weather cooperates, many young turkeys will make it to their first fall.

SECTION III

Getting Ready: Guns, Bows, Garb and Gear

Choosing a new gobbler gun and determining a preferred load is one of many off-season tasks.

CHAPTER 7

GUNS, BOWS & AMMO

Selections Have Never Been Better

The first shot I ever had at a wild turkey was with a 12-gauge, double-barreled Ithaca, the same gun into which I loaded No. 6 shells for rabbits and ringnecks and "punkin balls" for use in the deer woods.

It was a matter of using what I had, which wasn't much for a country kid whose sole arsenal consisted of one shotgun and a single-shot .22 for squirrels.

I missed that lone hen and didn't get the opportunity at another fall bird for several more seasons. By then I'd picked up an old 16-gauge Remington semi-automatic for $50 – a serious investment in the early-60s on a first-year school teacher's $74 per week take-home pay.

Who could then have imagined that 30 years later turkey hunting's popularity would spawn an array of shotguns with quality, features and cost to satisfy everyone?

The double-barreled shotgun was once the most popular all-purpose firearm, as shown in this 1947 photo of Dr. Bennett of Centre County and his jake.

As with turkey calls, turkey guns are largely a matter of personal "fit," ranging from 20 gauge magnums for small-frame shooters to 10- or 12-gauge shoulder-busters whose patterns with 3-1/2 inch shells can reach out and tumble a turkey at 60 yards.

Given an educated guess, it's my bet that the most widely used firearm for turkeys is the 12-gauge pump or semi-automatic fitted with full chokes and capable of shooting 3-inch magnum shells. Twenty-two to 26 inch barrels are the most desired.

I won't even venture an opinion as to which manufacturer produces the "best" turkey gun. Arguing about Winchesters, Remingtons, Mossbergs, New Englands,

The Remington Model 870 SP Magnum pump remains one of the most popular turkey guns.

H&Rs, Brownings, Benellies, Ithacas, Rugers, Weatherbys or Berettas is fodder for campfire discussions.

For the record, guns I've used to kill my last 10 turkeys included the semi-automatic Remington 1100 SP, the pump Mossberg Ulti-Mag 835, single shot, 12-gauge Thompson Center Encore and most recently a Remington 870 Mag – a light, sweet-handling 20-guage that my wife has her eyes on for next season.

Of course, you'll continue to see a good number of turkey hunters afield who carry the same guns – largely pumps and double-barrels – that they use for small game. But if you have your mind set on a scattergun made especially for turkeys, the following suggestions may be of help.

Your Next Turkey Gun

The first consideration when buying a turkey gun is fit and feel. Shouldering and cheeking a gun, safely swinging it to an object on the wall and placing your head in a position to clearly and quickly focus down the barrel is the first order of business. Next, consider the comfort level. Is it sufficiently light to haul around the woods for six hours? Can it be held in position for more than a minute or two? Can you comfortably adjust your forward hand up and down the fore-stock? Can you easily and quietly flick the safety with your thumb? And what about the gun's kick? Is it manageable?

Answer yes to each question and you're a step closer to finding the gun you'll need.

Don't believe that because you're a big man that you need a big gun. If you want a shotgun that will maintain a reasonably tight pellet pattern at 50 or 60 yards, then invest (and quality firearms are true

Bob Long of Fountain Hill made good on a 62-yard shot with his Benelli Super Black Eagle II.

investments) in a hard-hitting, long range firearm.

But if your self-imposed limit on shooting is 30-35 yards, like most others, you can safely reject the old "bigger the better" concept. With today's screw-in chokes and magnum loads, packing enough wallop isn't the problem it was two or three decades back.

"Back then," turkey guns were not designed for small frame men, women and kids. Today you'll encounter a respectable variety of choices, from semi-automatics to single shots. The latter is of special importance when choosing a gun for a youngster. Whether with or without hammers, single-shots are safer, simpler in function and of lower cost than most other guns. You'll only get one shot, of course, but youngsters seldom have the composure or skill to fire a second round at a spooked gobbler. That shouldn't be a compelling concern.

Set Your Sights

Several other items also need attention before plunking down your hard-earned money. A functional sighting system – if it's not already factory-supplied – is the second order of business. Many multi-purpose shotguns are sold with single-bead sights, arguably the leading cause of misses. Shooters who do not cheek the stock properly may believe their aims are true, but often they're looking over the lone bead near the muzzle while watching the target's neck and head. Such shots typically are high and will contribute to a gobbler's master's degree when a load of No. 6s zoom by its crown.

Nearly all of today's turkey guns are equipped with mid-barrel beads which line up with the bead near the top of the barrel. Aligning both sights forces the shooter to cheek the stock properly and line up the beads.

My choice is the colorful fiber-optic sight system mounted front and back on vented-rib and plain barrel shotguns. Not only are they adjustable for elevation and windage but they provide for quick and easy sighting in both sunlight and dim conditions. Fiber-optic sights include a bright green rear V-sight and a red or orange front sight. More and more firearms manufacturers are including fiber-optics on their turkey guns.

Yet another option is a telescopic sight, to which a growing number of hunters are switching. Don't rely on just any old scope discarded from

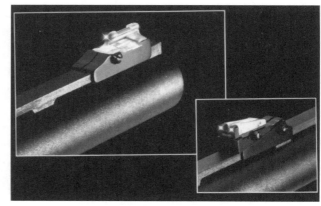

Fiber-optic sights, glowing green and red or orange light, are adjustable front and back.

your deer rifle, however. Rather, check out the red-dot or cross-hair scopes, some of the latter with standard reticles and a large circle or diamond for centering on a gobbler's head. Local sporting goods stores and, in particular, catalog companies offer a variety of scopes specifically for turkey guns. Anyone with vision problems should consider a scope of no more than 4x, preferably the weaker 2x to reduce the chance of misjudging distances as a bird works within range.

Tom Neumann of Penn's Woods Calls is one of a growing number of turkey hunters switching to scopes for their shotguns.

All Choked Up

Shotgun chokes are surely nothing new but now they're so commonly used by turkey hunters that many new guns are sold with removable, screw-in chokes. The devices constrict the shot, making for tighter patterns and consistency at all distances. Some models are ported for less recoil. Chokes are not universal and must be purchased for specific barrels.

Camouflage

Another item demanding attention is camouflage; not necessarily a must on a turkey gun but a sure benefit. The glint of a blued barrel in the morning sun has alerted more than one turkey to the presence of trouble. Modern guns come with tough, dipped barrels and stocks but expect to pay a bit more. An option is to cover the shiny spots or the entire gun with camouflage tape or fabric camo sleeves that fit over the stock and barrel.

Swivels and Slings

Most turkey guns come with factory installed swivels. A gunsmith can easily attach them if they're lacking. Many modern guns also come with slings. My preference on all shotguns and rifles is to purchase wide, neoprene slings such as offered by Butler Creek and others. They cushion and grip the shoulder better than the nylon webbing or fabric slings. Consider also that the use of a sling makes for safer gun handling when walking.

Shotgun Ammo Choices Galore

It wasn't all that long ago when spring gobbler hunters firmly believed that "the larger the shot size, the better the chance for bagging a longbeard."

Some hunters utilized No. 2 or bigger shot and a few took a chance with buckshot. After all, they reasoned, No. 6 shot is used on pheasants and No. 4s on ducks. A turkey is 10 times bigger so a heavier load is mandatory.

That was then – when most shotshell selections were not designed for turkey hunting.

This is now – when shotshell manufacturers are catering to the nation's nearly three million-plus turkey aficionados with a stunning array of shells made for modern day firearms.

Now smaller shot size is more the rule than the exception. Consider that shotshells with small pellets also hold more pellets, thereby providing denser and more consistent patterns at 20-40 yards than ever before. Of course, larger (but fewer) pellets provide more penetration. As said, it's a personal choice.

Many states, Pennsylvania included, have regulations barring large shot sizes. Here, only No. 4 or smaller pellets are legal. Safety considerations were at the core of the mandate.

That pretty much limits shot size choices for turkey hunting to number 4, 5 and 6 loads, although some hunters are going to 7-1/2 shot for maximum pellet numbers. Today, according to Alan Corzine, director of research and development for Winchester Ammunition, No. 6 is the most widely used shot for turkey hunting.

"It simply patterns better" said Corzine, with whom I shared a room on a recent hunt. "But that doesn't necessarily

This gobbler was taken at 22 yards with a tight-patterning 20-gauge Remington 870 loaded with 3-inch, No. 6 Remington Hevi-Shot.

Tight-patterning chokes are recommended for most turkey guns.

mean No. 6 will perform better in everyone's gun."

As in muzzleloading, the proper pairing of a barrel, choke and load is essential to accuracy and maximum pellet penetration. Even guns exactly alike may provide different patterns.

Truth be known, your favorite turkey gun will probably pattern satisfactorily with any number of the many shotshells available, so you don't have to buy a costly selection before patterning. But it is wise to experiment with two or three types of shells and determine for yourself which is your most effective load.

The Turkey Rifle

I met a fellow at an Allentown outdoors show last January who handed me what he called his "turkey hunting pictures." Each was of himself and a good supply of fresh meat for the freezer. Indeed, the photos included a turkey. They also showed a pheasant, two rabbits and two squirrels, all taken on opening day of the season.

Truth is, in autumn many rifle-carrying hunters go "free-lancing" and sometimes run into flocks of hens and their young or are granted a welcome encounter with a gobbler. When I asked the gent what he used I fully expected him to tell me it was a .22 Mag, a popular turkey rifle.

It wasn't. His firearm was a .17 Remington, which has acquired a following in the past few years and is arguably as close to an all-purpose mini-load as anything else. When properly placed, the tiny 25-grain bullet will drop a turkey with little, if any, destruction of meat.

So, too, is the .22 Mag, a favorite at ranges of fewer than 100 yards. Add to that the .223 and .22-250 Remingtons with longer reaches. The .243 and 6mm also draw

Carol Hoats and her husband Scott of Danielsville sight in a .22 Mag, a popular fall turkey rifle.

some fans but chances of "blitzing" a bird comes with the higher calibers.

That boils down to accuracy. A hit in the head – as with shotguns – won't spoil any meat nor will the bird run off. A body shot, however, may damage the breast no matter what caliber is used and the injured prey may run or fly off and die. The head is a small target at 50-100 yards, which drives some riflemen to aim for the base of the neck instead of the standard head-or-neck placement.

A quality, variable scope should be standard equipment for long distance shooting. More accurate aim is possible with a high-resolution optic that serves to double the benefits with light-gathering performance on cloudy mornings.

Finally, it should go without saying that line-of-fire mishaps are more likely to occur with rifles than with shotguns. Not only will an errant projectile travel farther than pellets but the potential is there for greater injury should another hunter be "in the line of fire."

Blackpowder Birds

As serious turkey hunting shifts from being a mystique to that of a science, a small core of hunters is going afield seeking an extra challenge. Instead of carrying shotguns capable of decking a gobbler at 50 or 60 yards, they're making it a close-up game, as have many deer hunters across the years.

Hunting turkeys with a muzzleloader is largely driven by nostalgia and history. Buckskinners who haul out their traditional double-barrel guns and load them with shot rather than bullets for turkey hunting relish the idea of seeing white smoke between them and incoming gobblers

Call-maker Robby Rohm of Perry County bagged this gobbler with a Knight TK-2000 muzzleloading shotgun.

when the trigger is squeezed.

Not all muzzleloading firearms are of the traditional design, of course, as Knight Rifles, CVA, Thompson-Center, Traditions and others each produce a "modernized" shotgun or two. Although the latter may have some current features such as fiber-optic sights and screw-in chokes, they function every bit the same as traditional firearms.

Both types of front-loaders work as well as modern shotguns at limited distances. Most muzzleloader hunters will use No. 4, 5 or 6 lead shot and an equal volume of black powder or Pyrodex. As in patterning modern guns, a pre-season session with turkey targets is in order, testing two or three loads in the process and determining maximum ranges of each.

Ken Merkel of New Smithville used his Double Bull blind to take four gobblers in four successive springs.

The same "possibles" used for deer hunting serve double duty on turkey guns. Wads, shot cups and lead spread-shot are about the only items not also used for whitetails.

Of course, no matter if you're after fans or antlers, cleanliness of the blackpowder firearm is a necessity.

Bows and Blinds

The legalization of blinds for turkey hunting in 2001 is having an impact on diehard archers looking to take gobblers with bows-and-arrows.

It wasn't long ago that bowhunting for turkeys was considered more a stunt than an ethical pursuit. Making the draw when a turkey was in range or trying to hold a draw for too long a time as a strutting bird or two closed in made for taking bad shots. If the broadhead pierced a vital zone, great. If not, a body shot may have injured the bird and off he went down a hill, running or gliding, only to die elsewhere.

By using a blind, however, a bowhunter's movements are hidden. Those who now carry their "hides" on hunts, especially on field edge set-ups, enjoy closer and more accurate shots at birds that have no idea they're targets.

Although I've not hunted turkeys from blinds, I have photographed from them and have witnessed first-hand even a wary gobbler's acceptance of a blind, sometimes moving as close as 10 yards to the set-up.

While all lightweight blinds have potential for turkey hunting use, those with black interiors to hide the archer's movements are preferable. Blinds made of thin fabrics allow light to pass through and lighten the interior – and your movements. Heavier camouflaged fabrics are good choices, however. "Windows" through which to aim should also be abundant and quick and quiet to open and close.

Oh yes, a couple decoys set about 10-12 yards from the blind will help by distracting any incoming birds and doubling as a distance reference.

Dr. Greg Caldwell of Port Matilda called in and bow-shot his gobbler during the fall season.

Bows and Broadheads

As for archery gear, requirements are much as they are for deer hunting. The best choice is the largest broadhead available and legal. You can also get into a quick argument by recommending mechanical broadheads (which open upon impact) but some bowhunting turkey authorities will highly recommend them. It's an individual choice, of course.

An innovative newcomer on the broadhead scene is the 125-grain Gobbler Guillotine, an oversize, four-

The new Gobbler Guillotine broadhead, designed for turkeys, is proving itself as an effective alternative for head/neck shots.

blade broadhead measuring 4-inches wide that lives up to its name of Guillotine. While you must still make an accurate shot, this giant-size broadhead accommodates a margin of error that makes the jugular easier to sever. On precise neck shots it can actually decapitate the bird. Successful turkey archers suggest that the bow be sighted for 15 yard shots. While it may not be pretty to watch, fewer bow-shot turkeys will

An archer's aim point is determined by a gobbler's body position.

escape and go unfound.

The blade requires (and comes with) longer shafts so that the Guillotine does not come in contact with the arrow rest when drawn.

Invented by Texas archer Matthew Futtere, the blade was not a subject of scrutiny by any game agencies as this is being written in late 2004. Said Futtere: "I advise anyone using it to first check with their state wildlife agency." For more information, log onto: WWW.ARROW-DYNAMIC-SOLUTIONS.COM.

As for the bow, the same one used for deer is just fine. However, you may want to crank the draw weight down a few pounds in anticipation of a need to draw and hold for a long period, until a bird gets into the desired distance and position.

Where to Aim

Unlike aiming a shotgun, when the turkey's head may be in any position to take a safe shot, archers have specific areas in which to place their arrows. One – if the bird's looking toward you – is the standard head/neck area. This, of course, requires precise shot placement.

In this situation, most archers will shoot for the upper breast just above the beard. If the target is standing broadside, aim either for the head and neck or the wing butt, which is in line with the vitals. If the gobbler's fanning and looking the opposite direction, shoot two or three inches above the anus. Anything striking high should hit the spine.

If you're serious about bowhunting, a life-size 3-D turkey target should be added to your list of suggested Christmas or birthday gifts.

MIKE BLEECH
Warren, PA

Mike Bleech is a well-known Pennsylvania outdoor writer, book author and columnist for the Warren Times Observer. He hunts in Warren, Forest, McKean, Erie and Crawford counties.

CHOOSE A CHOKE TO FIT YOUR HUNTING STYLE

Every serious turkey hunter should pattern-test a shotgun before hunting with it. Not only do you need to know where the center of the pattern is, you also need to know how large the pattern is at various distances. That includes testing each choke tube using the exact load used while hunting.

Those super-tight choke tubes might be great at longer distances, but for close shots they are far too tight. Small patterns make it difficult to hit a moving turkey head or neck, especially when you might be shooting from an awkward position. Tighter choke tubes certainly have their place. Use one if you have to call a bird more than 40 yards without cover, of if you are hunting a bird you know has demonstrated previously that it will not come inside 40 yards.

I expect that a modified choke tube might be best for most Pennsylvania situations. Hunters should choke for the shooting situations they anticipate and carry at least a couple extra tubes so they can be switched as the specific situation requires. Be sure you know what these chokes will do by preseason testing.

Shooting situations do not always occur as they are anticipated, of course. This is one of the reasons I hunt with an over-under shotgun. Often I will install a skeet II choke tube in my bottom barrel and a full choke tube in my top barrel. This keeps me in reasonably good shape for shots from 15 yards all the way to 40 yards.

Another important factor to consider comes down to the moment of truth as the bird comes in. By making the proper choke choice, you'll have the memorable opportunity to watch a gobbler strut in rather than having to shoot at the first opportunity.

DENNIS STRAWBRIDGE
Dallastown, PA
Dennis Strawbridge is serving his 20th
year on the P-NWTF board of directors.
In the off-season he makes collectible
turkey calls of exotic woods, terrapin
shells and wingbones. Dennis hunts
Perry, Wyoming and Potter counties.

BASIC BEAD SIGHTS
WILL DO THE JOB

I prefer to use the standard bead sights available on all shotguns, along with a ventilated-rib. Many times I have considered other options such as rifle-type sights, scopes, electronic red dots and the popular tritium and other light-gathering sights.

I choose bead sights as they are capable of serving my purpose as long as proper shooting form is maintained. The second bead at mid-point is a very important aspect of this system. By aligning both beads you are assured you will not shoot over your target, offsetting any shooting form errors you have introduced while hunkered down in the awkward shooting positions.

I also like the fact that I can see my surroundings for maximum safety. When a follow-up shot is required, standard sights serve the best opportunity to make the shot, especially if a bird has taken flight.

Following are stories that some of my friends and associates have experienced that help justify my choice of sights for my turkey gun.

Several years ago one of the people in camp was using an electronic red-dot sight that was carefully adjusted in the dark to ensure the sight would be ready. The next day, when a gobbler got into range, to the dismay of the hunter the red dot could not be seen in the full daylight.

The hunter sighted thru the device, placed the bead of the barrel on the bird's head and proceeded to take the shot. Unfortunately this placed the shot pattern in a tree behind the bird at about the height of a standing person. With that, the bird was gone.

The second incident involved a scope. One morning I called in a jake for a friend. He allowed it to come really close as he had never killed a turkey. To our dismay, the bird took wing at the shot and was gone.

When he got home, he patterned the gun and found it was shooting low and left about a foot. He had sighted the gun with slugs the previous fall and never patterned it for shot before spring.

DAVE EHRIG
Mertztown, PA

Dave Ehrig, dubbed "Mr. Blackpowder" by his fellow traditionalists, is under contract to write an authoritative "Muzzleloading for Deer and Turkeys" to be published in 2005. He seeks gobblers with his frontloader in Berks and Bradford counties.

MUZZLELOADING LOADS: VOLUME FOR VOLUME

The sun had just lifted over the horizon as a loud gobble echoed down through the hollow. I cautiously set up my Buckwing jake and hen decoys and slipped into the mental mood of the hunt. My daydreaming, however, had caused me to miss the fly down of the birds.

To my advantage, a small flock had left its treetop roost a mere 50 yards away. Minutes later I caught sight of periscoping red and gray heads skirting the wheat field, just 25 yards in front of me.

Ka-boom!

The air filled with a cloud of white smoke.

By the time the sound echoed off the far hill I was admiring a fine 19-pound eastern wild turkey flopping at the edge of the wheat field. While smoke from the 100 grains of FFg Scheutzen black powder flavored the morning mist with sulfur, the equal volume of No. 5 chilled lead shot did its job. The thoroughly modern Traditions Turkey Pro muzzleloader topped by a Bushnell Trophy 1.75x4 scope had delivered a dense pattern at 30 yards.

Hefting the big bird to my shoulder, I couldn't help but wonder what that first buckskinner must have felt like as he sighed in relief with the promise of food from his smoothbore Trade Gun.

So you want to give blackpowder turkeys a try and are unsure of what load to use? Remember the words "volume for volume." It signifies the volume of powder compared to the volume of shot, a one-to-one ratio. For instance, I feed my old T/C New Englander sidehammer percussion 85 grains of FFg powder, followed by a vegetable-oiled, fiber wad over the powder, a Cabela's plastic shot cup sliced twice to create petals, and 1-3/8 ounces of No. 6 c cylinder bore pattern at 25 yards. With the shot cup, the pattern tightens to a definite modified-choke pattern.

After shooting the T/C New Englander, it was immediately obvious that the extra-full Remington choke made a difference. Thirty yard patterns were impressive and even the 40-yarders were lethal. By reducing the diameter of the bore near the muzzle, the shot string was compressed and elongated, creating a large percentage of pellets on target.

Even with the 100 grain load of FFg and an equal amount of No. 5 shot, using a red Ballistic Products shot cup, the balanced volume of powder and shot produced the best pattern.

Nickel-plated, copper-plated and chilled-lead shot were all tested. In spite of their higher price, the nickel and copper plated shots didn't do better in patterning and in some tests performed worse than the chilled lead. While eye-popping, we weren't sure why that happened and chose not to use it.

While either shotgun – the Traditions in-line or more primitive T/C New Englander – could have taken the Eastern turkey, the load would still have been the same.

◆ ANOTHER VIEW ◆

KEN MERKEL
New Smithville, PA

Ken Merkel runs an archery shop in southeastern Pennsylvania and has been pursuing turkeys for 20 years. In 2004 he shot a 17-pound tom – his fourth in four years with a bow – holding a 10-1/2 inch beard. Ken hunts Lehigh and Schuylkill counties.

BLINDS AN ADDED DRAW
FOR BOWHUNTING

I started bowhunting in 1963 as a preliminary for rifle season. However by the mid 80s I was totally addicted to archery.

I tried bowhunting turkeys for a few years but I was always detected when drawing my bow. Out of frustration I just quit hunting them.

Then, in 2001, about three weeks before the spring season opened, Pennsylvania legalized hunting from a blind. I felt reborn. I immediately ordered a Double Bull Pro Staff T5 blind. On the second day I used it I found myself at full draw on a gobbler only 12 yards away. He was mine. I followed that up with three more longbeards in the next three spring seasons. I'd like to share some of my findings with you.

First and foremost is the blind. I prefer the Double Bull over any other because it has no wind flap and can be set up in about 20 seconds. I always set up two or three decoys where I want to take my shot.

I've found it best to leave only two shooting windows open to keep it dark inside the blind. I hunt only one or two locations in an entire morning and call only enough to get them coming. I've learned to be patient.

I hunt with a 60-pound compound bow shooting Gold Tip arrows and NAP Gobbler Getter broadheads. I prefer a sight with one moveable pin because the birds will come right into the decoys if you don't spook them with careless movements or noise. If they do hang up, I can determine distances with a range finder. That takes out the guessing game and gives me the exact yardage.

For blind hunting my choice is a dark camo. My favorite is the original Trebark. I wear a glove on my bow hand because it gets close to the window.

On broadside or quartering shots, I will shoot for the thighs. Take out their legs and they can't run. I've never seen one get airborne that didn't have full use of his legs. I will shoot for the base of the beard on incoming birds. Birds with tails erect and facing away at full strut give you a perfect bull's-eye.

All my birds succumbed within 20 yards but if one gets out of sight wait as long as possible to go after him. He won't go far. If you bump him too quickly you'll push him out of the area with a very scant blood trail. I practice shooting on my knees and sitting on a stool.

I hope these tips will help you to successfully bowhunt turkeys. It will add a new element to this cherished spring challenge.

❖ ANOTHER VIEW ❖

MIKE LAMADE
Pleasant Mount, PA
Mike Lamade is the executive director of the United Bowhunters of Pennsylvania. He writes for Bowhunter magazine and gives seminars at sports shows each winter. Mike hunts in Wayne County and across the Delaware River in Sullivan County, New York.

LONGBEARDS THE HARD WAY: GRAB YOUR BOW

His whole body trembled as he extended his lust-filled swollen neck.
Gobble! Gobble! Gobble!
Drooping his wings and spreading his fan, he sucked his head and neck down into his shoulders and ruffled his body feathers forward. Simultaneously, he inflated himself with air, taking on the appearance of a big black ball. As his

contracting muscles forced the air from his body a guttural b-o-o-r-r-m! sent vibrations into the solitude of the spring morning. I found myself looking directly into the haughty, mahogany eye of the most magnificent gamebird in North America.

That scenario occurred almost 40 years ago, but my excitement for turkey hunting has never diminished. It's actually gotten stronger since I've put down the shotgun and began hunting these critters with the bow. I've been fortunate to have bowhunted most of North America and have been to Africa eight times, but taking a gobbler with bow-and-arrow is right up there with the harvest of a moose or kudu.

Hoping to share the same with others, here are a few tips for anyone considering bowhunting for gobblers.

• Use the deer-hunting bow with which you are most familiar. Crank it down a few pounds as you may be at full draw for quite a while. Total camo is essential if you're not hunting from a blind. Movement is your greatest enemy.

• Your deer broadheads are fine, but shot placement is the critical issue, not the broadhead. Flying turkeys don't leave blood trails.

• On profile shots, aim for the wing butt. Better yet, wait until he goes into full strut and shoot him at the base of the fan when he faces away from you.

• Blinds and decoys work great! I've used a Double Bull blind the past three years and have done very well. Wear black clothing in the blind and the birds absolutely won't see you, even when they're as close as 10 yards.

• Consider the new "Gobbler Guillotine" broadhead. It has a 4x4-inch cutting width and is designed for head/neck shots at close range. It's truly devastating. He's down in an instant, or you completely miss. If you try the Guillotine, be sure to follow the manufacturer's instructions for blade installation, shafts and fletchings. They are critical. If you have trouble finding them at your local shop, Cabela's and Gander Mountain both carry them.

Give hunting longbeards "the hard way" a try. You'll love it!

CHAPTER 8

DRESS FOR SUCCESS

Outfit Yourself for Fashion and Function

It's hard to believe that quality camouflage is a relative fledgling in the hunting marketplace. In the early 1980s most deer and turkey hunters were still trying to hide from their prey wearing traditional camo in generic military and tiger-stripe patterns. Only one pattern exclusively for hunters – Jim Crumley's Trebark – was on the market and it took a few years for sportsmen to take full notice.

But when they did, a new market was born and it was to give a colossal economic boost to the hunting industry. Bill Jordan introduced his Realtree wear in 1986 followed by Mossy Oak and the parade was under way. Throughout the remainder or the '80s and into 2000 and beyond camo choices have taken the hunting world by storm.

Blending in with the environment is the role of modern camouflage, such as this hunter dressed in an ASAT Ultimate 3-D suit.

Some four dozen or more companies have come (and some have gone) with several no longer making their own garb but licensing their patterns to clothing manufacturers and for use on everything from boots to box calls. Each year these companies offer variations on their own themes.

Check magazine and TV ads, the racks in sporting goods stores or the pages of the many catalogs coming to your door and you'll find such familiar names as Realtree, Mossy Oak, Trebark, Tru-Woods, ASAT, Bushlan, Predator, Diamondback, Rockaflage, Skyline, Timber Ghost and more. How long the camo revolution and evolution will last is surely a subject on the minds of most manufacturers. Just how many more new yet effective designs can they create?

Choosing a pattern is largely a personal matter, influenced by the dominant habitat in which a hunter travels and the varied styles and incidentals – from turkey vests to gloves, hats, boots and rainwear – available.

Form and Fashion

Of course, function in hiding a visitor to the woods is the key ingredient in camo choice, but fashion also plays a role in the industry. Visit a sports show or turkey hunting seminar and you'll see sport shirts, sweaters, vests, slacks, dresses and other attractive camo garments not meant to be worn in the woods.

Fabrics run the gamut with late-season gobbler hunters yielding to thinner, cotton materials but, in many states, necessarily opting for fleece, wool and insulated garments on cold mornings.

To suggest an "ideal" outfit for hunting in Pennsylvania would be futile and we won't even attempt it here. Admittedly, it would largely depend on personal choice and what I like won't necessarily be appealing to you.

Another big

When properly cared for, today's camo holds its patterns and colors for many years.

consideration is how much money's in the checking account.

It's arguably safe to say that camouflage is more important for turkey hunters than deer hunters, although bowhunters may question that claim. The big difference is that deer are largely color blind and see in varied shades of black, gray and white. Turkeys, however, have acute color vision, as do all birds. Unlike deer, turkeys have poor night and dim light vision. That's why they spend their nights in trees and seldom – except for being blown out of bed by a storm – is night vision necessary.

Camo Head to Toe

For turkey hunting I wear colors to match the drab, late April-early May woodlands – grays, browns, bronzes and blacks primarily. As the season progresses I'll switch to similarly patterned pant, jacket and vest with the addition of green to mimic the greening woods. At times I'll also don mismatched camo – one pattern on the bottom and another up top to aid in breaking up my human outline.

When you've selected your garb, search out related accessories including gloves, caps, hats and face masks to match your camo. The flash of a white hand or face is enough to alert an incoming bird.

It may go without saying, but no matter what your camo choices they will only perform their job of blending you with your immediate habitat if you stay still. Scratch your nose or move your head with a gobbler watching and it won't matter much how effectively your camouflage performs.

Only careless movement will give away this hunter calling from a rock outcrop.

Best Foot Forward

Turkey hunters tread in places that would quickly make most normal people turn around and head home. Picture a dawn greeted by the eerie howling of coyotes across the Pocono Mountains, timber rattlers lying in wait on a rocky Tioga County hillside, black bears watching out for their cubs and bobcats slinking through the hazy dawn in the Susquehannock seeking a kill.

So what is it that ranks near the top in the "big hurt" department among hen and gobbler gunners?

How about blisters?

It's a good bet that more turkey hunters have been slowed or temporarily put out of business by battling blisters created by wet and untested boots than any other affliction – including rattlesnake bites.

On a cool, spring morning when the toms are gobbling something as seemingly minor as a blister, pinched toe or an abrasion on your heel can quickly spoil the day, one step at a time. By the time you get back home or to camp for some much-needed first aid, the problem may have become one of downright misery.

As with most afflictions, prevention is the key. It begins with choosing the right boots and properly breaking them in long before setting foot in the woods.

Buying Boots That Fit

When purchasing footwear, "try on both boots and fully lace them as you would when setting out on a hunt," advises Bill Johnson of Georgia Sport & Trail Boots. "If the fit is bothersome, try on another pair in the same or a different size."

As everyone's feet are different shapes and sizes, it's impossible to create a boot that's perfect for everyone. Seams stitched into the boot lining may produce "lumps" here or there which are sometimes not detectable until the boots are worn for a time. If you feel the seams when you first slip on a boot, it will surely become more of a problem later.

Properly fitted, waterproof boots, such as this one from Georgia Sport & Trail, prevent blisters and provide traction and comfort on all types of terrain.

The difference from one boot size to another is surprisingly small.

"A half-size is a mere one-twelfth of an inch around your foot," Johnson explains. "It's the same difference between width sizes – such as size D to size E."

If boots are bought locally, slip on the pair and walk around the store, standing tiptoe, balancing on your heels and even walking while over-shifting your weight to the left or right. Tightness or your feet sliding front and back or side-to-side somewhat simulates what will occur in the field.

Break all boots in by wearing them to go to the store, mow the lawn or take your morning walk.

Some turkey hunters wear rubber boots, which are also prone to cause blisters when wet. Hunters who know they'll be crossing streams or swampy areas will often opt for knee-high rubber boots, which are now available in several camo designs. They serve double duty when walking across high-grass, dew-drenched fields in the early morning.

Sock it to Me

When boot-shopping, take along a pair of the socks you will be wearing while hunting and slip them on before starting your "test-drive."

Drying wet boots this close to a heat source, such as a woodstove, is not recommended.

Depending on your choice of hunting hosiery, you may also find the need to invest in a couple pairs of hunting socks.

"My pet peeve is the guy who spends $100 on a pair of boots, then puts on a pair of cotton socks," says Joe Hanley of Rocky Boots. "Around here we have a saying; "Cotton is rotten," he laughs. "Cotton absorbs sweat and traps it on the foot and there's no better way to get blisters."

A thin polypropylene or acrylic sock liner and a Thermax, wool or wool-blended sock worn in a Gore-Tex boot is ideal. Liners wick perspiration away from the feet, through the outer sock and eventually through the boot via Gore-Tex. In turn, the foot is kept dry and the chief cause of blisters is eliminated.

A change of socks in midday can work wonders on long walks in turkey country.

I always pack an extra pair of socks for changing in midday; the dry hosiery providing a fresh step to my travels.

Let it Rain

My first piece of hunting rainwear, bought way back in the mid 1950s, was a $2.95 plastic "waterproof" parka that was ripped from my body first time out as I shuffled through thickets and spiny fencerows.

Fast forward 50 years to the era of Gore-Tex and Dry Plus, Cordura and Ripstop, Realtree and Mossy Oak, Thinsulate and Slimtech – and other familiar trademarks. Staying dry and keeping the wind at bay is today achieved with little more than a paper-thin liner and tough outer shells tender to the touch.

Who'd ever have thought it was possible?

The silent, waterproof, windproof, "invisible" rainwear of today isn't typically bought on a whim. Now rainwear is an investment in comfort and reliability and, properly cared for, can last nearly a lifetime. What you use in the turkey seasons is just as likely to serve the purpose during the deer seasons.

Consider these tidbits of advice before laying down your credit card or cash.

• Start by slipping into a jacket or parka and move about, lifting your arms as if shouldering a gun or drawing a bow. Make special note of how quiet (or noisy)

Frogg Toggs Rainwear weighs but 11-1/2 ounces and is easy to carry in a vest.

the garment may be. Remember also that you'll be wearing raingear over other garments, depending on the day's temperature range, and choose your size accordingly.

• Familiarize yourself with product functions of such items as Gore-Tex, Air Flow, Heat Tech and many others which are described in catalog copy or explained on a garment's hang tags.

• Check the locations and sizes of the pockets, including special pockets designed for carrying walkie-talkies and GPS units or stowing a flashlight, knife or game calls. Also check the pockets on the pants, such as the side cargo pockets. Note whether the pockets are open at the top or protected by storm flaps secured by snaps, buttons or zippers?

• Most rain jackets or shells have hoods, which typically aren't utilized until rain begins to fall. In between raindrops, however, they can be annoying. Some rain jackets have hoods that can be rolled up and zipped into the back collar or completely removed and carried in the coat.

• A must on any wet weather outfit is a vertically-zippered pant bottom on the outside of each lower leg which opens to permit slippage over boots. Also look for snaps on the pant cuffs to adjust for short- or long-legged wearers.

Check out all advertised features, such as oversize belt loops, zippered air vents or jacket-pant combos that can be rolled up and stored in their own side pockets or in small sacs. The easier it is to carry raingear in your turkey vest, the more likely you'll take it along on those "iffy" days when it may or may not rain.

Don't stay in bed when things look dreary. Turkeys don't stay inside when the weather turns foul.

Neither should you.

Cabela's Millennium Rainwear is built for use in both cold and wet weather.

ERIC MILLER
Stowe, PA

Eric Miller is president and designer of Pennsylvania-made Tru-Woods Camouflage. His best gobbler ever is a 22-pound Potter County tom with a 10-inch beard. In addition to Potter, he patrols the woodlands of Berks and Montgomery counties.

BE CREATIVE IN YOUR CAMO CHOICES

With due respect for the visual acuity of the wild turkey, the changing landscape across a May season presents just as much of a challenge for the hunter. Depending on where you choose to hunt, you'll need to pick a camouflage outfit to fit that area.

I recall an opening day a few years ago in Potter County. The date was April 30th and there wasn't a hint of green in the north woods. Scouting weeks earlier, I located lots of fresh gobbler droppings as well some wing feathers. I was pumped to say the least, but I also became aware that a camo pattern with green in it would not do well in concealing me from a lovesick gobbler.

The Potter County woods were still leafless, except for the stubborn taupe leaves of the American beech saplings. I knew a camo pattern that utilized more of the darker browns and tans would be ideal. I opted for my Tru-Woods Autumn pattern and it allowed me to blend in perfectly with the fall-like setting and ultimately led to tagging a nice gobbler that opening morning.

As the season progresses, changing your camouflage selection to correlate with the changing leaf growth makes good sense. As the woods green up, select a pattern that matches the colors of the leaves. Remember, turkeys do see color and matching your surroundings gives you another edge in your quest to put your tag on a longbeard.

When scouting, always take note of what the foliage is doing. If the green leaves are dominant, choose a predominantly green pattern. It makes good sense. In some instances, all of the leaves on the trees are green but the forest floor is still void of most signs of spring at which time you may opt to wear pants with a darker camo pattern to blend in with the browns and tans of fallen leaves. Wear a pattern with more green for your shirt or jacket, facemask and hat. I consistently do this with the Tru-Woods Autumn and EF/S (early fall/spring) patterns.

Now if you're a hunter on a budget or one who just prefers to wear one pattern, most camouflage companies offer patterns dubbed "multi-functional camo." Multi-function patterns like my Tru-Woods Transition 4D utilize both browns and greens throughout the pattern, allowing you to blend in with a wider variety of foliage. By placing just the right amount of both browns and greens in the pattern, the surrounding colors of the area you are hunting will actually pull out the appropriate colors to match.

Simply put, when you're in foliage that is browner, the browns will be more noticeable in your camo pattern. When you're set up in green leaves, the green colors are more apparent.

Regardless of what you choose to wear, be sure to match up with your hunting area and you'll be well on your way to blending in with your environment.

CHAPTER 9

GOBBLEROPTICS

Binoculars & Spotting Scopes for the "Birdwatcher"

Birdwatching takes on a new dimension for millions of outdoors people each spring. While enjoying the warblers migrating north along bottomland hardwoods and necklaces of geese in the warming sky, a select few camo-clad "birdwatchers" relish most the appearance of a long-bearded gobbler 30 paces off the front bead of a shotgun.

It's always puzzled me that a deer hunter wouldn't be caught dead without a binocular. Yet, over the years, I've seen more turkey hunters afield without the optic than I've seen with them.

Of course, optics aren't needed to identify a turkey, even at distances greater

Scoping distant turkeys in power line or gas line cuts reveals birds that might not otherwise be seen.

than 30 yards, but hunters who carry them know binoculars can be every bit as useful in the turkey woods as in the deer woods. The magnifiers offer the hunter a method for studying flocks feeding in openings several hundred yards off, for scanning treetops in the dim morning light, identifying far off hens from toms and as safety aids.

Cautious movement in set-up positions permits glassing and identifying incoming yet out-of-range birds.

The "light-gathering" binoculars I tote on spring and fall mornings are the same ones I carry on many deer hunts. They hold large-diameter objective (front) lenses which gather light, enhancing the user's ability to decipher detail under dim conditions.

Full Size or Compact?

My preference is for full-size 8x or 10x "gobbleroptics," but on any given day I may carry a smaller and lighter compact model. Compacts inherently have smaller objective lenses and do not gather light under dim conditions as does larger glass. But the size is appealing.

Cosmetics also play a role in binocular choices, especially for the turkey hunter. Most manufacturers offer them in a variety of camo patterns as well as the standard black or green.

Protective coatings are important on the lenses. By virtue of the places hunters, sneak, peek, creep, crawl, walk, run and slither, any turkey hunting optic will be exposed to dirt and moisture that won't be encountered at a NASCAR race or on a bird watch. Understand that a rubberized – often called "armored" – binocular finish doesn't guarantee they're water-proof. Don't confuse the term with "water-resistant." It's not the same thing.

Water-resistant models are typically coated with a tough rubber that repels moisture and offers protection should it slam against a rock or tumble off the truck's dashboard. Water-proof optics are just that. Drop them in a stream or soak them during a downpour and they'll continue to function without fogging or damage to the internal lenses. Plan on paying a bit more for fully water-proof binoculars but also accept that a fogged lens on a "bargain" binocular renders it useless.

Yet another type of coating deserves the shopper's attention. Modern glasses are covered with reflection-reducing material, such as magnesium fluoride, which enhances brightness. Be aware also that quality (in other words, higher priced) binoculars will have coatings on all lens surfaces. Low-cost optics may only have external lens coatings.

Spotting Scopes

Although not nearly as popular with turkey hunters as with big game hunters, spotting scopes are showing up on the spring and fall turkey hunting scene. These specialty scopes are available in fixed powers ranging from 15X to 45X, ideal for making ultra-long distance sightings. Zoom lenses typically in the 20X-45X range are also available. Front lenses are interchangeable on some scopes.

Several years ago on a hunt in Berks County my partner and I scanned a vast greenfield with binoculars and, some 500 yards off, saw what appeared to be little more than black dots. It was a flock of turkeys feeding along a field edge. It was obvious through my 10X binocular that two of the birds were strutting, hinting that they were longbeards.

I'd stuffed a 20X Simmons spotting scope and a miniature tripod in my vest's gamebag that morning and we immediately rested it on a deadfall to study the birds "close up." After a couple minutes we concluded that both of the fanned tails were on the hind ends of jakes.

We studied the birds for another 30 minutes to see if a boss gobbler was among them or would wander into the field to claim the harem. But when none appeared, we moved on to search for other birds. Had we decided to move close for a set-up, not knowing that the strutting birds were yearlings, we'd have wasted another hour or so of valuable time.

Tripod mounted spotting scopes can save steps and time when used to find and identify turkeys in wide, open spaces.

Rangefinders

Hunters who use several types of optics may want to use a rangefinder when set-up time permits. Although I use rangefinders for deer hunts and for pinpointing big game 100 to more than 500 yards away, I seldom use them for gobblers. But I can't deny that they are useful for turkey hunters who need to know precise distances, revealed by scanning and recording distances of several obvious objects such as a large tree, rocks, deadfall or even decoys when there's sufficient time available during the set-up.

Scoping Other Hunters

Yet another advantage of carrying optics – specifically binoculars – is for the identification of fellow callers in the woods. No longer is this facet of turkey hunting incidental. The number of turkey hunting related injuries and fatalities, although decreasing in Pennsylvania, surely suggests the need for more precautions while afield.

"Shot in mistake for game" is the reason cited in most of the reports. Such unforgivable accidents occur when spur-of-the-moment decisions to shoot are made by hunters who have not identified their targets. Binoculars strung from the neck and readily available for use can prevent such misidentification.

Defensive hunting is also enhanced with optics. A hunter set up against a fat oak may glimpse a movement through the woods but not be able to determine if it's a turkey, a deer or a human. A closer view with binoculars will aid in identification and making the decision on whether to "sit tight" or move on.

Binoculars and scopes have always been standard gear for big game hunters. Now they've moved to the turkey woods where they're as important a part of spring "birdwatching" as calls, camo and common sense.

Make a "gobbleroptic" a standard take-along item on every spring or fall quest.

Optics played a role in this successful hunt.

**BOB
WALKER
Media, PA**

Bob Walker is known nationally in shooting and hunting circles as the inventor of the Walker's Game Ear. Although he hunts turkeys across the country, his favorite retreat is his log hunting camp in Sullivan County.

"BINOCULARS" FOR YOUR EARS

Second only to vision, the most valuable sense we employ when turkey hunting is our hearing. Let's run through a typical turkey hunting day to demonstrate this.

The night before the hunt you await sundown and try to "put the birds to bed." You move from place to place and offer barred owl calls in an attempt to get a bird to gobble.

If you're fortunate enough to find an occupied roost, or a single longbeard, you return early the next morning and set up on him. You make some soft calls while he is on the roost in hope that you will hear a gobble.

You get a response and your heart shifts into overdrive. He's double and triple gobbling on the limb and as the sun colors the eastern sky he pitches down, hits the ground and shuts up.

You wait him out for a time, then ask yourself: "Did I spook him? Does he have hens with him? Is he gone?"

About then you hear something, a squirrel, perhaps. You slowly peer around the tree against which you're sitting. Your eyes quickly focus on a gobbler – and his on you.

You've been busted.

If only you'd heard the bird walking in the leaves sooner, it may have been a different outcome.

Many people, young and old, have hearing problems. Natural deterioration, noisy work environments and neglect take their toll over the years. The most hazardous recreational sport affecting hearing is shooting. Whether you are hunting or at the range, it is of utmost importance to wear hearing protection. Few people wear plugs or muffs when hunting, which is understandable.

In 1989 I came up with the idea of what came to be known as the Walker's Game Ear – called "binoculars for the ears" by one outdoor writer. This hearing enhancement and ear protection device was designed specifically to help hear game better and protect hearing at the same time.

My training as a Pennsylvania-licensed hearing health-care specialist and my affinity for hunting all aided in the Game Ear's design. Watching hunters and shooters who were not wearing hearing protection helped me to understand what was needed to fill this necessary niche in hunter safety.

Each time a shooter discharges a firearm the eardrums suffer. The unique aspect of the Game Ear is that it both amplifies and protects. It amplifies high frequency sounds, the kind made by most game.

I don't mean to sound boastful but I am proud to not only have made an aid to help hunters hear when in the woods but can also be used for protection when trap or skeet shooting.

Whether it's a double-duty Game Ear or muffs, plugs or some other protective device, remember to wear something to prevent hearing loss when enjoying your time on the range and afield.

CHAPTER 10

VESTED INTEREST
Unique Items for Toting Afield

I am convinced that if someone made a turkey vest with 75 pockets, I'd probably fill them all. Turkey hunters are like that. The preseason preparation for getting things in order requires that some items simply be left at home or you'll feel like a bag lady toting stuff around the woods that you don't really need.

The following isn't a dissertation on how to purchase or pack a vest, however.

Turkey decoys have a come a long way from the early 1980s when Harry Boyer of Newport fashioned a crude but effective hen decoy with a foam silhouette embellished with actual turkey feathers

It's no problem today buying a quality vest enabling you to carry a variety of items – including dead turkeys. Rather, this chapter covers the various odds and ends that are useful enough to make you want to stuff them in a pocket and try them out under actual hunting conditions. Some may never be used again while others, like decoys, will become necessities.

Fake Jakes and Hens

I don't know who made and successfully used the first turkey decoy but I suspect it was someone like my friend Harry Boyer of Newport who fashioned a foam form of a turkey silhouette and glued feathers on it, then painted the head a blue gray and shoved a wood or metal support through the belly area. It worked.

Glenn Lindaman shows the effectiveness of his BuckWing decoys on a Lehigh County farm.

Later came the Feather Flex decoys followed by Flambeau, Sceery, Renz, Delta, Carry-Lite, H.S. Strut and others. A few years ago my friend and hunting companion Glenn Lindaman added a line of decoys to his BuckWing Products out of Whitehall, Lehigh County that have made big news because of their lifelike qualities. Fittingly called Lifelite Collapsible Decoys, they're as close to the real thing as I've seen. Included in the line are the standard hens and jakes along with a strutting jake and a breeding pair.

In 2004, Glenn went a step further with his Bobb'n Head creation which, with the slightest breeze, moves its head as if feeding. All decoys are displayed on an expander which is inserted inside the shell and opened like a small umbrella to bring out the full width of the imitation for rapid placement.

Strangely, some hunters frown on decoy use, which is surely their privilege. Most hunters just won't leave home without them. Newcomers and archers, in particular, will have more success in bringing birds close in open field situations with decoy set-ups. The visual lures are so realistic that boss longbeards may attempt to mate with the hens or beat up on the fake jakes.

Given the choice of carrying but one decoy in my vest's gamebag, I'd choose the jake. Usually, however, I'll take a pair of hens and a jake if I plan to spend the morning in open woods or field edges.

Oh yes, when on the move and the dekes are placed inside the game pouch, cover the red heads of all jake decoys with an orange sack or ribbon for obvious safety reasons.

Unusual Calls

So you thought you'd heard it all?

You haven't. And neither have those tough gobblers.

Some innovative calls may add new twists to the standard yelps, purrs, cackles and putts when the big boy quits answering or simply bellows his head off at 50 yards but refuses to take another step.

That's when you may want to spit, drum, scratch or squeeze.

A jake is instrumental in any decoy set-up.

A gobbler declares his victory upon flattening a hen decoy.

A gobbler's strange drumming sound, which some hunters have difficulty hearing, can be heard for 100 yards or more by other springtime hunters.

The Spit-N-Drum from M.A.D. Calls and Lohman has been around for several years. It never got real popular – except for the savvy hunters who learned to use it and may now be hard to find.

According to Mark Drury of M.A.D., who developed the one-of-a-kind call several years ago, the idea for the Spit-'N-Drum hinged on the fact that most hunters have heard a gobbler's pfffft-rummmm when strutting, so they know what it should sound like. I've hunted with partners who could hear gobblers drumming at 100-150 yards and more under optimum conditions. I can't hear it for more than 20 yards or so.

Turkeys, of course, can hear these subdued drumming sounds quite well, which is the idea behind the unique call. The sharp "spitting" sound (sometimes called the pulmonic puff) precedes the "drum," a low, droning sound. It's believed both sounds have similar origins, created by vibrating the breast muscles and forcing an expulsion of air via the pulmonary system.

Oh yes, some hunters with mimicking talents have learned to make the sound without the aid of an instrument.

Wing It

This non-vocal enticer is used when turkeys fly down from the roost in early morning. It duplicates the rhythm of flapping wings because that's what it is – an actual wing.

Some wing calls are fashioned from manmade materials but the Primos

The Real Wing mimics the fly-down wingbeats produced by turkeys leaving their roosts.

Game Calls' contribution – called The Real Wing – is a cupped wing of a domestic turkey dyed to resemble that of a wild bird. Its curvature and stiffness provides both maximum "thumping" sounds and enables easy storage and transport in a vest. An easy-grip handle holds the feathers in position.

"Obviously, the number one use for this is as a fly-down call," said Will Primos, who first thought of the wing ploy when he noted roosted gobblers flying toward the thumping wingbeats of hens descending from roost trees more than 200 yards away.

Lohman's version is dubbed The Wing Thing. The compact, 6x6-inch,

Every hunter should have a call or two that functions under rainy conditions. The Woods-Wise Wet Friction Calls do just that.

camouflage, heavy-fiber device contains vents which, when slapped against the hand or leg, create the fly-down, fly-up, fighting or flapping sounds.

Wet Weather? No Problem

Forget about foul weather when using the Woods-Wise Mystic Wet Friction Calls series of friction calls, says Gary Sefton of Woods-Wise Products.

"We've developed a new surface material that creates consistent friction without the need for chalk," Sefton explained. "It's permanent, consistent and maintenance free. No slips, skips or errors no matter how wet they get."

This isn't a brand new concept but the calls are new – and they work. Sefton and company owner Jerry Peterson have come up with a variety of waterproof designs – from box and push-pin calls to slate and glass calls (all coated with a product called Greenite). Even the strikers – dubbed "wet sticks" – are coated.

Every turkey hunter should have at least one of these in his or her vest for use on both mild days and, in particular, whenever the skies yield rain.

Gobbler's Lounge

A dozen or so years ago Bucklick Creek introduced its internal-frame support turkey vest (You don't need a tree to lean against. Flip open the frames and sit down anywhere – quickly.). I used mine until someone decided to "borrow" it from the side porch of a Mississippi turkey camp.

Now I have the Gobbler's Lounge which fulfills a similar function – and more. Made by Little Big Horn Outdoors, the vest holds an internal frame for back support in a seated position. Foam seat

The Little Bighorn Gobbler's Lounge Vest has a built in back support for those quick set-ups when no suitable tree is available.

and back pads provide additional comfort when against a tree or holding your ground in a field situation where no suitable back support is nearby. Side straps are adjustable and angled pockets provide easy access for calls and whatever else you're toting. A water-proof game bag offers additional storage spaces and serves as a carrier after you've bagged your bird.

It's available in Realtree Advantage and Mossy Oak Break-Up. A smaller Jake Lounge for young or small-framed hunters is also available.

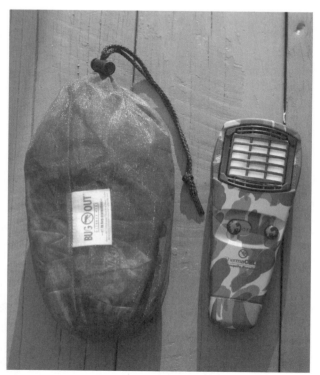

The Bug-Out Bug Pack and ThermaCELL Mosquito Repellent readily fit in vests and will make those "buggy" spring mornings tolerable.

Keeping the Bugs Away

Ever have a day in the woods when the insects are relentless– especially mosquitoes, no-see-ums and biting flies? Believe me, storms of biting insects all trying to puncture your hide or sip your blood will affect your concentration and movements, which turkey hunters cannot afford.

For the past half dozen years and more I've made certain I take my mesh Bug-Out Bug Pack – pant, jacket and head net – in sizes S-4XL. The chemical free suit, weighing mere ounces, rolls up to the size of a third of a loaf of bread and is instantly available when things get "buggy" on warm days. The mesh jacket in Mossy Oak Break-Up or Advantage Timber can also be worn over a dark T-shirt on hot days.

A bit more hi-tech is the odorless ThermaCELL Mosquito Repellent, a non-battery operated device that you simply lay at your side when you're set up. About the size of a TV remote control, its repelling scent is effective for 225 square feet around you. It operates on a butane cartridge which heats a 1x2 inch mat of pyrethrin, a naturally occurring insecticide. I tried it and it works.

If the woods are alive with the annoying hum of bugs buzzing your head, a ThermaCELL will take care of the problem.

The Streamlight Stylus flashlight throws a powerful and concise green beam for working your way through the woods before first light.

Let There be Light

No matter if you're hunting deer or turkeys, it's often dark enough to need a flashlight when you work your way into a set-up site long before sun-up. Last gobbler season I was introduced to the Streamlight Stylus, a ballpoint-pen-size device with a small but brilliant green beam that can light your way without scaring the game. As I moved toward a roost site I had to make a cut through a pine woods with a soft, quiet bed of needles but loud-snapping twigs underfoot. A regular flashlight would have revealed my presence to the treed birds.

But the directional Stylus' oval beam enabled me to see the debris and avoid the noisy, dead twigs. As it was pointed downward there was no glare directly from the light that could be seen by anyone but me.

Streamlight also makes the ClipMate which, as the name suggests, clips to a hat or jacket when hands-free use is necessary. The tough aircraft aluminum lights are each powered by a 60-hour battery and an LED bulb which last for 100,000 hours, which isn't likely to expire before we do.

Hand Pruners

Yet another "don't leave home without it" device is a pruning shear, the same one you use for trimming around the house. If you're shopping for one specifically marketed to hunters, consider a shear with a ratchet mode for easier cutting.

Hand pruners come in handy for trimming twigs and low brush at a set-up site.

The device comes in handy when trimming brush from the base of the tree against which you plan to set up or when trimming twigs and small seedlings that could interfere with shooting.

Hunter's Ground Seat

The first time I saw a metal-framed hunting seat it was hanging from a long strap looped over Rob Keck's shoulder as we made our way into an Alabama turkey woods. I couldn't imagine how such a device wouldn't constantly get snagged on brush and low limbs. But when Rob let me try it out a bit later I was convinced I needed one. And I haven't gone turkey hunting without it ever since.

The legged seats on newer models have foldable and adjustable legs along with the standard nylon webbing that is considerably more comfortable on the butt than the cushioned seats hanging from most turkey vests. Several companies market them. The more comfortable you are during a prolonged set-up, the less you'll need to stretch muscles and move.

I don't leave home without it.

Sing a Song of Turkeys

One last item isn't carried afield but will surely get you in the hunting mood on the way to the turkey woods. It's a CD of songs titled "The Magnificent Bird: Songs About Wild Turkeys and Wild Turkey Hunters" written and performed by Gary Sefton of Primos.

I've spent time in turkey camps with Sefton and each evening his guitar-strumming and singing "All Henned Up and Happy" or "Turkey Talking Blues" among others was the perfect ending – and can be the welcome beginning – to a day in the woods. For more information, call 1-800-735-8182.

Of course, these aren't the only items carried in my turkey vest. A knife, toilet paper, calls, raingear, sandpaper and chalk, gloves, head net, binoculars, small zip-lock bags, water, sandwich, candy bars, small digital camera, camo hat and ….. Uh-oh. I think I just ran out of room.

Growing in popularity among turkey hunters are nylon-webbed, lightweight seats such as the one being used by Randy Rice of Lancaster.

SECTION IV
Turkeys from Forest to Farm

Pennsylvania offers more than 4 million acres of public forests open to hunting.

CHAPTER 11

A WELCOME CHANGE BORDER TO BORDER

More Turkeys and More Places to Hunt Them

The welcome change is more rule than exception across the Keystone State. What welcome change?

The presence of turkeys, of course, along with the places they're today being hunted. Some counties have maintained healthy numbers of turkeys since the 70s and 80s when forest-game species began replacing upland game on the hunter's most-wanted list turning a surprisingly vast amount of Penn's Woods' forests, fields, farmlands and suburban woodlots into turkey havens.

How do turkey numbers in Pennsylvania compare to, say, those of 30 or 40 years ago?

"It's as good as or better than it's ever been, "said Arnie Hayden without hesitation during a tape recorded interview in 2001," even though there are a lot more hunters after them, especially in the spring."

"We believe the population minimum – notice I said minimum – today (2001) is about 150,000 birds but as many as 200,000 may be a closer number," said Hayden. "Remember that hunters are shooting them now in places that didn't

Trap-and-transfer birds of the 1980s established turkey populations in the Northwest.

even have them 15 or 20 years ago."

Along with the spring sport's growth, thanks largely to plenty of press, seminars and videos erasing some of the mystique of hunting and calling elusive gobblers, hunters are today pursuing "backyard" turkeys all over the state.

I asked Hayden for thumbnail profiles of the wild turkey's status across Pennsylvania, which has remained stable in recent years.

Northwest

Consider the strand along the Ohio border in Mercer, Crawford and Erie counties as a prime example of turkey restoration programs meeting with success. Few turkeys were found here 25-30 years ago but prolific trap-and-transfer birds have now established themselves in swamp bottoms and adjacent forests. Fall hunting was closed here for a time to allow the birds full protection and the opportunity to breed and extend their ranges.

Some hunters and biologists go so far as to rank the general Northwest as the best region in the state. Here turkey numbers are healthy and hunting pressure somewhat lighter than in the remainder of the Commonwealth.

Southwest

Fall hunting was banned for a time in portions of the Southwest but since then the region has come on strong.

"There are more turkeys here now than in some counties that have always had turkeys present," said Hayden. "Hunters here no longer have to travel north to find them. They can stay right at home."

He listed Greene and Washington counties as showing the greatest success, noting that progeny of wild birds translocated there in 1981 and 1982 have "really come on strong." Even heavily-populated Allegheny County surrounding Pittsburgh holds birds, although nearly all hunting requires landowner permission.

The region is typified by farmland punctuated by wooded valleys and numerous woodlots. The bellows of gobblers are now heard across the region's towns and villages as the birds have adapted to heavy human habitation. Somerset and Greene counties are arguably the region's best bets. It's not new turkey terrain though. Flocks here have been healthy and hefty for many years, said Hayden.

Northcentral

The vast Northcentral region has always supported good turkey numbers.]

"There's not much to say about this region that hasn't been said," declared Hayden. "It's always been good and still is. Much of our trap-and-transfer stock came from here."

Pressed for specifics, he listed McKean, Elk, Cameron and Potter as bright spots while recalling the years of and following the devastating winter of 1978-1979. Hayden did some studies immediately thereafter, finding an exceptionally high toll on turkeys, deer and other wildlife. It took a long time for flocks to recover as thousands of them succumbed to the incessant cold, snow and freezing conditions. Of course, other regions also suffered the memorable (for those of us old enough to remember) winter.

The big woods country of the northcentral range has a broad selection of public hunting grounds, portions of which are so remote that few hunters put forth the effort to penetrate the rugged wilderness where largely unhunted turkeys are found.

Southcentral

Here turkey numbers have been relatively stable. The first-ever trap-and-transfer birds were released in Perry County in the 1960s and hunting there has been good ever since. However, the growth of turkey numbers accounted for what Hayden described as "an over-exploitation" in the late 60s. A bright spot is that there's no problem finding public hunting grounds with hens and toms in any of the counties.

Hayden chose Huntingdon and Bedford counties as tops in the region with

The Raystown Lake area in Huntingdon County is one of numerous public hunting lands in the southcentral region.

Adams County's trap-and-transfer birds slowly taking hold. Recent trap-and-transfer operations (from 2001 to present) have been focused on Lancaster and York counties. Research continues on the failure of traditional turkey hotspots in the Michaux State Forest in WMU 7B, including portions of Cumberland, Adams and Franklin counties continues.

Northeast

This is a good news-bad news region.

"Bradford and Susquehanna counties (bordering New York) are the best when it comes to turkey numbers and hold the sort of landscape allowing them to expand their range," biologist Hayden said. "Both those counties have excellent habitat diversity. It's not all unbroken tracts of forest like a lot of other places."

Wayne County is also coming on strong but populations in Pike and Monroe counties may have dwindled over their former numbers. Yet pockets of turkey concentrations, such as found near the Delaware River, particularly the Delaware Water Gap National Recreation Area, offer excellent prospects for tying a tag to a gobbler's leg.

Non-contiguous tracts of Delaware State Forest and numerous gamelands permit easy access for hunters seeking public hunting grounds.

Southeast

This is town-and-farm country with gamelands scattered across the farming landscape south of the Appalachian Ridge. Turkeys have set up housekeeping everywhere, making a knock on the door necessary for hunting private lands. Attribute that to the successful trap-and-transfer projects of the 80s and a few more recent stockings.

Not all of the Southeastern turkeys are as suburbanized as this one photographed by Barry Miles atop his Easton home.

Hayden ranked Schuylkill as "far and above" the best county in the region. Lebanon, Dauphin and the hilly sectors of Lancaster and Berks also host growing turkey numbers. Here farmland and large woodland tracts set the stage for the gamebirds' growth.

Turkeys are also in residence near the Philadelphia suburbs in Chester, Bucks and Montgomery counties but large, private holdings prevent all but the chosen few from attempting to call them to gun. The Lehigh Valley – primarily Northampton and Lehigh counties – has also come on strong as a turkey haven thanks to numerous woodlots in which to roost and plenty of farm crops to get them through winter. The fragmented landscape, however, limits the birds' natural distribution.

Given a choice of one section, however, the vast Blue Mountain ridge of the Appalachians stretching from the Susquehanna River eastward to the Lehigh and Delaware rivers offers a near continuous strand of state gamelands. Fall hunting is not permitted in Wildlife Management Units 5A and 5B in the extreme southeast.

CHAPTER 12

PENNSYLVANIA FORESTS, PARKS & GAMELANDS

More Than Four Million Acres of Trees and Turkeys

Every spring and fall about a quarter-million Pennsylvania hunters head to the woods and fields with plans for calling and culling hens and bearded gobblers.

Each year some 70,000 or more succeed.

Even though many of the state's wild turkeys have taken up residence in farmland and near-suburban areas, many hunters "head to the hills" for their May and November pursuits. The "hills," more properly classed as mountains, are typically in the state's northern tier and the southcentral counties where deep-woods habitat interspersed with scattered farm and pasture lands offer the birds food, shelter and nesting grounds.

But some these sizable public hunting lands have an identity problem.

"Got one up on the gamelands near Promised Land early this morning," a camouflage-adorned hunter told me when I queried him on his success between sips of coffee at a Pike County diner last spring.

Further quizzing revealed that he wasn't on gamelands at all; rather on a Delaware State Forest tract, one of 20 State Forest District holdings scattered across the commonwealth. It underscored my belief that "gamelands" has come to generically identify any forested tract open to hunting.

Both out-of-state and resident hunters seeking new terrain on which to ply their turkey calling talents should pay particular attention to Pennsylvania's array of forests and gamelands, along with some federal holdings, shown on official state road maps. The public lands are easily located and highly accessible.

State Forest Lands

It's the "other" 2.2 million acres of wooded, state-managed landscape that is frequently confused with gamelands. The 20 state forest tracts are owned and managed by the Pennsylvania Dept. of Conservation and Natural Resources Bureau of Forestry. The Game Commission and DCNR foresters regularly work together in management programs for everything from grouse and woodcock to turkeys and whitetails.

"People have a tendency to consider any forests open to hunting as state

Trails and gated roads provide ready access on state forest lands such as Delaware State Forest in the Poconos.

gamelands," said Dan Devlin, a forestry department wildlife biologist. "Our studies show that about half of all state forest recreational use is by hunters."

Much of that is made up of turkey hunters.

Devlin said management programs on State Forest Lands have enhanced turkey production in the past couple decades. Successful trap-and-transfer programs have initiated new flocks or bolstered low numbers in places that historically held turkeys.

Devlin points to the northcentral region, specifically Sproul, Susquehanna, Elk and Tioga state forests as holding consistent turkey populations.

"That region historically held good turkey numbers and it still does," said Devlin. "All of the state forests there have good access and hunters can distribute themselves pretty well."

The bureau is involved in a continuing project with the Pennsylvania Chapter of the National Wild Turkey Federation creating herbaceous openings planted with legumes, clover, crabapples, fescue and other food-producing vegetation in Michaux S.F. and 11 other districts.

Devlin said it's difficult to list "hotspots" as most of the state-owned forests have viable turkey populations.

"Better that a hunter should first scout some of the forests," advised Devlin.

He recommends looking for sign in places that have been recently logged and now offer abundant, low-growing plants.

The openings appeal to turkeys for food and nesting.

The well-maintained roads – more than 2,600 miles throughout the system – are all open to vehicles during most hunting seasons. Turkey hunters often cruise the gravel roads and broadcast owl, crow or hen calls

Recent clearcuts on both gamelands and state forests provide prime nesting and feeding habitat.

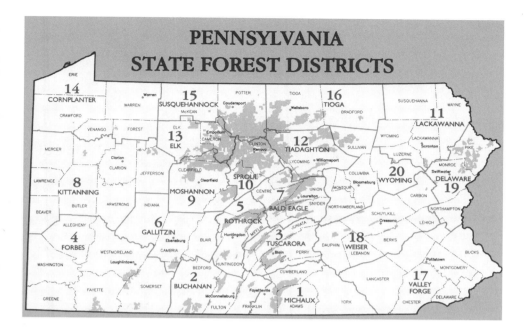

PENNSYLVANIA
STATE FOREST DISTRICTS

to locate turkeys on the ground or ready to roost.

Of big help in navigating state forest hunting lands is a series of maps and guides published by the forestry bureau. They show both state forests and nearby gamelands plus gravel and logging roads, hiking trails and other landmarks in each region. Most of the tracts within each district are not contiguous.

For information and map requests, call 717-783-7941.

State Parks

The Bureau of Forestry's sister agency – the Bureau of State Parks – also owns nearly a quarter-million acres, much of it surrounded by state forest land or otherwise also open to hunting. Many of

Many state parks offer both camping and hunting.

the 114 parklands are located in close proximity to state forests and offer both camping and hunting.

Campsites vary from primitive to modern and rental cabins are also available at some parks. For details, call 1-888-PA-PARKS and for additional forest/park information log onto www.dcnr.state.pa.us.

State Game Lands

Unlike state parks and forests which were purchased with public tax money, the 1.4 million acres of gamelands scattered across the state were bought with funds derived from hunting license sales and other land-related sources, such as timber and gas sales.

Today land conservancies across the state help create gamelands by raising funds to make up the difference between the sale price and the $400 per acre limit imposed by state law.

The locations of the 300 or more gamelands tracts, all identified by numbers, are shown on state highway maps. For more detailed information on each tract, log on to the Game Commission Web site at www.pgc.state.pa.us.

A map of each gameland and its

Turkey hunters using game and forest lands in Elk and Cameron counties may cross paths with elk.

location is shown with municipal boundaries, roads, waterways, impoundments, trails, shooting ranges and more. Click on Hunting and Trapping, then State Game Lands and the county of interest. Maps can be printed for carrying afield.

Allegheny National Forest

The largest tract of public hunting grounds in Pennsylvania lies within Allegheny National Forest in Forest and Warren counties, on the New York border. The federal woodland's 512,000 acres cover portions of Warren and McKean counties in the north and Elk and Forest to the south. Numerous roads, both black-topped and unpaved routes, penetrate the public hunting grounds.

The national forest also contains the Allegheny Reservoir (also known as the Kinzua Dam) which offers afternoon fishing when turkey hunting is over for the day.

A few years back I spent a spring morning hiking the landscape within the national forest with Dave Streb of Quaker Boy Game Calls. The well-known caller lives just across the state line in New York and the close proximity makes him a regular on the national forest scene.

While that day we didn't return with gobblers draped from our shoulders, we enjoyed exclusive use of several square miles of the territory during the heart of the season. We didn't cross paths with a single hunter.

"You can expect some pressure on Saturdays but during the week you pretty much can have things to yourself here," said Streb. "Carry a compass and a topo map because this is big country and you don't want to get lost here."

Adventurous hunters can boat to non-pressured shorelines and set up turkey camp and several backwater campsites exist along with many other campgrounds.

Despite first glances hinting that the rugged mountains may not be worth the climb, outdoor writer Mike Bleech, who lives near the famed national forest, says otherwise.

"The area is known for its rugged hills but this is actually a plateau. The terrain is relatively mild if you stay away from the major streams. Since you will most likely find gobblers near the tops of the hills you can save a lot of climbing if you start from roads that cross the tops of the hills."

Otherwise, gobbler hunting in this mountain-country differs little from the calling and set-ups in other parts of the state. However, the many miles of secluded dirt and blacktop roads winding through valleys and across mountaintops offer hunters the chance to "cruise" – stopping and calling every half-mile or so in hopes of getting a response.

"The locals do a lot of that and it doesn't take very long into the season to educate the birds," says Streb. "Last year I was working a bird when I heard a truck stop. Someone got out and called and the bird stopped gobbling. I heard the truck leave and 10 minutes later the gobbling started all over again."

More than 65 million board feet of lumber is harvested from the forest each year, creating a diversity of early and second-growth habitat, ideal for turkeys and deer. Information on the locations of the recent cuts is available from the park office.

Particularly helpful for the first-timer "Hunting and Recreation in the Alleghenies," a detailed map, updated annually, showing roadways, timber cutting areas, and other tracts open to hunting in the region. It sells for under $2. For more information write: Allegheny National Forest, POB 847, Warren, PA 16365 or phone (814) 723-5150.

Delaware Water Gap N.R.A.

Yet another popular federal turkey hunting site is the Delaware Water Gap National Recreation Area on the Delaware River in Monroe and Pike counties. Under the jurisdiction of the National Park Service, the 70,000 acre facility (about 50 percent New Jersey land) offers some excellent turkey hunting.

Hunt here on May mornings and bring your fishing rod to cast for American shad in the afternoon. The migrants make their annual spawning runs at the same time gobblers are being pursued.

For information, call 570-588-2435.

BOB CLARK
Mechanicsburg, PA

Bob Clark is a past vice president of the P-NWTF chapter and a popular seminar speaker. He holds the honor of bagging New York's fifth-biggest-ever gobbler. He hunts Clarion, Cumberland, Elk, Monroe and Venango counties.

HUNTING PUBLIC LAND GOBBLERS

Being a hunter who treads mostly public lands, I frequent one of the most difficult-to-hunt state forests in Pennsylvania. The Michaux State Forest (Adams, Cumberland and other counties) has gone from abundant turkey populations for 20 years to a collapsed population the past 10. Hard winters, poor reproduction, predator population increases and hunter pressure have arguably made these wild turkeys the most difficult to hunt in all of America.

This may sound like an exaggeration until you consider that I have hunted turkeys in 45 states and killed gobblers in 39 of them. The tremendous growth and interest in turkey hunting has produced, in many areas, a hesitancy of the toms to gobble. Last year (2003) a very obvious lack of gobbling occurred during the second breeding period.

Blame, in part, the run and gun enthusiasm of leading callers across the country shown on television and videos. Many hunters have gotten the impression that this is the only way to hunt. Anyone who needed to be at work by 8 or 9 a.m. probably called too much, ran and pressured the birds too much and by these actions, educated the turkeys. The result was many spooked gobblers, including jakes, remained stone silent.

This situation in my own back yard, where my hunting lodge is located, gave me two choices; either change my hunting location or change my tactics. By spending many hours scouting prior to daylight and at sunset, I was able to locate gobblers sounding off. It was often one or two shorts bursts and then total silence.

I confirmed some observations with the late Ben Rogers Lee, my hunting buddy for more than 15 years, that the tree call, soft putts, and very slow or fast runs were the calls these birds delivered under pressure and stress. Ben taught me to listen carefully to identify these mature gobblers without ever seeing them. They need not gobble to reveal their presence. In mornings they become quiet but may get in the singing mood when hunters head to work. Today I'll often sleep until 9 or later instead of rushing to the woods at daybreak.

But if you need to lose an extra 30 minutes of sleep and get into a public hunting grounds early, especially in the spring, do it. If possible get farther into the woods where the birds haven't been disturbed. Many turkey hunters, like deer hunters, do not change their ways and will not hike into the backwoods.

Regardless of where you hunt, what is important is how you hunt. The volume and rhythm of several calls is important, as is unlimited patience. The ability to sit still and not move is as important as proper calling. Sometimes it's best to let the longbeards hunt you. Once I get their attention I never, ever call to or push them unless they communicate with me.

Get off the beaten path, particularly on hard-hunted public lands. Remember that you won't have to drag out a 150 pound deer.

By comparison, that 20-pound gobbler will be as light as, well, a feather.

DAVE AND RENEE MITCHELL
Fogelsville, PA

Dave Mitchell is a PGC land management supervisor in the state's Southeast. His most notable gobbler, shot with a bow in 2002, held 6-1/2 and 9-1/2 inch beards. He hunts in Lehigh, Northampton, Schuylkill and Carbon counties.

CHECK OUT GAME LAND TIMBER CUTS

The key to hunting turkeys is finding them and the best way to do that is to first find good habitat.

Some of the best habitat is in and around timber sales.

On State Game Lands, the Pennsylvania Game Commission harvests timber several different ways, each with its own goal.   For thinning to grow, our goal is crown development and spacing. In such a cut, we remove about 33 percent of the trees.

In a shelter wood cut, we plan for getting a future forest established by removing 50 percent of the trees.

In a regeneration cut, we take out 90 percent of the over-story to allow a future forest to take root. In all cuts we will save oak, hickory, beech, serviceberry, and black gum trees. All these trees produce seeds that wild turkeys utilize.

Timber sales usually need haul roads, skidder trails, and log landings. After completion of the sale, these areas are planted with a grass and clover mix. This will provide a forage area for turkeys as well as a site in which a hen can bring her brood to hunt for insects.

During the first few weeks of a poult's life, its diet is almost entirely made up of high protein insects, making the grassy feeding areas an important, new habitat.

We may also plant dogwood, sawtooth oak, chestnut and crabapple trees in and around the landings. This also encourages use of those areas as summer brood sites as well as fall/winter food sources.

If you're at a loss for a place to turkey hunt, check with your regional PGC office for the locations of any turkey-enticing cuts on State Game Lands tracts.

How will all this make you a better turkey hunter? In the spring, look for gobblers around clear-cuts. Hens like to nest in the thick cover provided by the downed treetops. If you find hens in the spring you also will find gobblers. The grassy roads, skid trails, and log landings make good strutting areas. In the fall look at areas that have been thinned or shelter wood cut. There should be an increase in mast and where you find food in the fall you will also find the turkeys.

The next time you are out on your favorite State Game Lands and see a timber harvest occurring, don't be upset. Good habitat opens the door for more game. You could be on your way to having better turkey hunting than ever before.

CHAPTER 13

TURKEYS IN THE STRAW

Farmlands and Suburbs the New Turkey Hotspots

Looking for a fat gobbler?

Head to the farms. Or suburbia.

That's' what a study of the changing world of Pennsylvania turkey hunting suggests. In the past quarter-century or longer the Pennsylvania Game Commission has reaped benefits from its turkey trap-and-transfer program, not only restoring gobblers and hens to forests from which they were eliminated at the turn of the century but also in farmlands and residential areas once considered inferior turkey habitat.

Yet, today they're back in surprisingly large numbers, nesting successfully and expanding their ranges throughout farmfield and woodlot country where 15-20

The author called in his 2003 spring gobbler on a lower Lehigh County farm.

Knocking on a farmhouse door prior to the season may lead to new opportunities almost anywhere in the state's agricultural region.

years ago the only gobbles came from pudgy barnyard butterballs. No longer must hunters living in farmland fringes head to the mountains for a chance at a gobbler. Today many stay at home for the grand spring hunt.

"We've had a lot of success in farm and residential areas," said biologist Arnie Hayden. "When we first started (trap-and-transfer) we never dreamt of going to farmland habitat for turkeys."

But following the success in restoring birds by moving them from northcentral and a few southcentral counties into other forest habitat, the agency decided to experiment with transplants to farmlands.

The process was first tried in Crawford County and the birds did well. Next came the release of truly-wild, not pen-raised, turkeys in Greene and Washington counties and later Beaver County, all in the far southwest. Pick one as tops today and Greene County will rank at or near the summit with its abundance of agricultural lands.

In the Northwest region, look to Erie and Crawford farmlands as gobbler hotspots. Each holds more than a dozen scattered gamelands tracts.

The highly-populated (with people) Southeast, with rolling hills and numerous woodlots checker-boarding farm and suburbs, yielded the biggest surprises for the success of translocated gobblers and hens.

"We once considered that region third class as far as turkey habitat," Hayden told me. "There are lots of farms and wooded areas, sure, but there's also a heck of a lot of people living there."

Today Bucks County is one of Pennsylvania turkey management's biggest

Natali Searl photographed this bold hen at the birdfeeder on the back deck of her upper Bucks County home.

success stories. Along with nearby Montgomery, Chester and Delaware counties which are highly suburbanized and Lehigh, Berks and Northampton counties, the region has become a stay-near-home turkey hunter's delight. All now have burgeoning turkey populations near and within residential areas. With portions of several WMUs closed to fall hunting as this is being written, the entire region has benefited from the ever-spreading flocks, the core of which were set free in 1987.

As with the Southeast's large deer herds, which find refuge behind No Trespassing signs, private places to hunt for turkeys is also a bit difficult – but not impossible – to find. Knocking on farmhouse doors or seeking out other landowners on whose lands turkeys prowl is the key to gaining access. Places such as Nockamixon State Park near Quakertown and gamelands surrounding the Middle Creek Wildlife Management Area of Lancaster County also hold healthy flocks and public hunting opportunities.

Today Bucks, bordered on the south by Philadelphia, ranks higher than some of the northern counties when measured by spring harvests per 100 square miles of woodland. As far back as the spring seasons of the early 1990s, Bucks County hunters bagged an average of more than two dozen gobblers per 100 wooded square miles.

On the other end of the state, Greene and Washington counties boasted phenomenal springtime gobbler harvests of 50.8 and 46.5. birds per 100 square forested miles, respectively.

Of course, farms aren't limited to only two or three sectors of the state. Cattle and crop lands can be found everywhere, including the valleys in big woods country of Tioga and Potter. Farther east, the state's dairy country stretches from eastern Tioga through Bradford, Susquehanna and into Wayne counties. The region offers a desirable mix of corn, wheat, alfalfa, oats, open pastures and surrounding forest – the best of both farm-and-forest worlds. Here, too, state forest, park and game lands can be easily found.

Transplanted turkeys have adapted well to farm country, surprising many biologists and hunters who see the turkey take hold in fields and edges where the ring-necked pheasant once reigned king.

Hunters in Wildlife Management Units 5C and 5D in the extreme Southeast are permitted spring hunting but not fall turkey hunting. Biologists believe any move to permit fall hunts would be counterproductive.

"The danger is in overharvesting," Hayden told me during our final interview.

"Only so many turkeys can be killed in the fall. Many of them will be young toms and it won't be long before the spring harvests drop as will the reproductive rates. Birds are vulnerable in those small woodlot areas where they stay throughout winter and in the brooding season. It's easy to overharvest them and a fall season could do that."

The grand spring (and fall) hunt in most places is no longer only a deep woods quest. Today camouflaged callers stand just as good a chance of tagging a longbeard by simply staying down on the farm – or locating places to hunt on the rim of the ever-spreading suburbs.

Field-edge set-ups are productive on spring mornings, especially later in the season when insects are present.

GREG NEUMANN
Delmont, PA

Greg Neumann and his brother Tom own Penn's Woods Game Calls and co-host the Hunting With Penn's Woods TV show. He has hunted a variety of counties but is most at home in Westmoreland. Greg's been pursuing longbeards for 26 years.

HUNTING FIELDS AND SMALL WOODLOTS

How do you change your mode of hunting in small woodlot habitats as compared to hunting the big woods?

Successful turkey hunting tactics vary slightly from region to region, depending upon the topography. Some of the age old tactics still apply, such as setting up above a bird or at least on the same level. Having a large tree to set up against and the sun behind you while keeping obstacles such as streams and fences from being between you and the bird are all important. But never set up in thick cover.

However, tactics do change somewhat when hunting "big woods" country vs. farm country with small woodlots. In the former, some of the previously mentioned set-up sites are far more critical. In the latter, the topography plays less of a role.

It is commonly known that turkeys like fields and open areas. Although they have excellent eyesight, they are even more comfortable in open areas where they have even better visibility. This is why setting up in an open area, whether clearings in big woods country or in a farm field, often proves very productive.

The use of decoys when hunting in farm country is most effective. Decoys set out in an open field provide high visibility to both an approaching turkey and to you, boosting your chances of success. It's my finding that the most effective set-up in farm country is with decoys placed 15-20 yards out from the tree line. You should be set up 5-10 yards inside the woods. This will give you plenty of cover to break up your outline so all attention from the approaching turkey will be on your decoys.

You can bet your best call that nearly every turkey in the area will be in the field sometime during the day. One thing to keep in mind when hunting open areas is that neither hens nor gobblers are fond of high, wet grass. If at all possible, try setting up in low-cut fields. Plowed fields are even better.

Another advantage to hunting woodlots in farm country is the ease of roosting a turkey in the evening, leading to a great set up in the morning. Turkeys will often spend the last hour of light in fields just prior to going to roost. Often times, these fields can be watched from far away without disturbing the normal routine of the turkey.

If you don't carry binoculars while turkey hunting, get into the habit of doing so. Glass the fields from far off so you know just where you want to be at the break of dawn.

CHAPTER 14

PEDAL UP A GOBBLER

Try a Bike Hike Into Turkey Country

A full moon hung low in the southwestern sky and dawn heralded first light on the eastern horizon as I slung my Mossberg over my back, mounted my bike and pedaled around the gate blocking motorized vehicles from entering a remote section of Delaware State Forest in Pennsylvania's Pocono Mountains. In the misty morning gray I rode 300 yards along the gravel road, then stopped to listen.

Not far off a distant whippoorwill greeted the dawn, its repetitive song piercing the Pike County woods. Soon a wood thrush joined in with its clear, flute-like melody. Above me a cardinal, always an early riser, broadcast its incessant "what-cheer, what-cheer" from atop a dimly lit maple.

The sound I wanted most

A biker-hunter can cover a lot of ground by accessing non-motorized roads with a bicycle.

to hear, however, was the wake-up call of a roosted longbeard. Drawing an owl hooter from my pocket, I issued an acceptable rendition of the "Who-cooks-for-you, Who-cooks-for-you-all" cadence made by barred owls, solid citizens of Pennsylvania's mountain country. The idea, as most turkey hunters know, is to stimulate a turkey to reveal its position, as they're known to do when owls, crows, pileated woodpeckers or even distant thunderclaps disturb them. But only a far-off barred owl returned the message and neither of us encouraged a gobbler to respond. I pushed off and coasted down the sloping road which meandered through a section of the state forest I came to know about 30 years back. For the past few years the gravel road I traveled had been closed to motorized vehicles, but bicycles are permitted.

Although it's said that one never forgets how to ride a bike, becoming proficient and comfortable with a model dramatically different from my one-speed Schwinn some 50 years back was made easier with practice runs in my driveway. Nevertheless, after several decades of not taking bike riding

Getting off the beaten path can yield largely unmolested gobblers.

seriously, I still found myself backpedaling rather than squeezing the handlebar levers when I wanted to brake. I knew I'd need to pay attention or end up in a ditch along the stony backwoods road.

Lance Armstrong I'm not

Pedaling another couple hundred yards along the darkened road, however, brought a feeling of some small confidence and when I shifted gears to climb a slope, I began to truly enjoy the swift, silent access provided by my newest toy. Indeed, my mode of transportation to the backwoods allowed me to hear the first gobble of the morning, no more than 175 yards off to the north. I braked, then followed with an owl call and received a second confirmation of the treed bird's location.

Stashing the camo-painted bike next to a deadfall, I quietly moved as close to the gobbler as I dared, then offered him a series of low-volume yelps, to which it gave an immediate answer.

Suffice it to say the next two hours were spent in a game of cat and mouse with the tom moving to within 50-60 yards on three occasions, only to retreat each time. The low greenery, typical of the Pike County forest, prevented me from seeing the bird, which probably would have been within sight earlier in the season when buds were just beginning to unfold and the forest floor was open. By 9 a.m., following a couple end runs, set-ups and calls, which each time brought the unseen but vocal tom close, he again retreated, distrusting the come hither hen talk without first seeing its maker. His last call, nearly three hours after the first, let me know he was on his way to the deep woods and wouldn't be returning.

I spent the remainder of the morning biking along other Delaware State Forest backcountry roadways, stopping now and again to call and listen, or simply focusing my attention in the treetops or the woodland floor for scarlet tanagers, warblers, thrushes and other migrants of the vernal season.

Unfortunately, as the morning progressed, winds increased and churned the greening trees, making it difficult to hear a response should a gobbler be enticed to answer the loud clucks, cuts and yelps from my box and mouth calls.

On one occasion I rounded a turn where a small buck, its small forked antlers covered in a milk chocolate brown velvet, and I surprised one another. It stood for several seconds before bounding off, pausing for a quick second glance, then melting into the forest's understory.

The bike that provided the refreshing experience and offered a new twist to turkey hunting is a Deep Woods Rider; a 21-speed Trek all-terrain prototype. Lightweight and easily stashed in the back of my Explorer, it's been a regular companion whenever and wherever feasible.

The bike's metal frame is covered in Mossy Oak Treestand camouflage applied in a unique process similar to that used on firearms and bows. However, despite my glowing report to the manufacturer, the bike never hit the marketplace. The reward for my testing efforts was a free (I know, nothing's really FREE!) bicycle.

I use the bike a couple times each season, putting on more mileage than I'd probably tread on foot and drawing stares from car-bound hunters I meet on some backcountry roads. My pleasant mornings spent pedaling past poplars and coasting through conifers has added a new dimension to the grand spring hunt, while providing swift and silent access to the Pocono Mountain backcountry.

Considering the hundreds, probably thousands, of state forest and state gameland roads – many of which are closed to motor vehicles during the hunting seasons – getting to spots where turkeys or deer haven't been harassed too badly has created a new twist to the grand spring sport of chasing gobblers hither and yon.

Give it a try sometime soon. First check with the regional state forest office for information on roads closed to motorized travel during the season.

It's a refreshing exercise and a great way to pedal up a gobbler.

Before the Hunt Preparation:
Safety, Patterning and Choosing Calls Top the List

Effective camo, safety orange and choosing the right calls help make successful seasons, as demonstrated by Rob Poorman of Muncy.

CHAPTER 15

CARELESS SHOOTING OR GETTING SHOT
It Could Happen To You

Ron Sandrus was reaching for his gun when the shots rang out.

Two rapid-fire reports from 35 or 40 yards off, the Hollidaysburg turkey hunter recalls. The next he remembers is being on his side, scrambling to get behind the tree against which his shotgun had been propped. The most poignant memory is the burning sensation in his left leg and the pain of a single pellet in the fleshy part of his hand, between the thumb and forefinger.

"I yelled but couldn't see anyone," Sandrus vividly recalls. "A minute or so after the shot I saw someone going over the ridge above me, maybe 100 yards away."

That was on April 25, 1992, the fourth day of the Pennsylvania spring gobbler season. Ironically, his then-fiancée Shirley Grenoble, a

Ron Sandrus of Hollidaysburg became the victim of a "mistaken for game" mishap.

well-known Pennsylvania turkey hunter and writer, was shot three years earlier not far from where Sandrus was peppered. Add to that Grenoble's and Sandrus' witnessing of yet another accident a year prior when Grenoble's son Mark was hit by pellets while the three were walking along a logging trail in Mississippi.

Ron has spent many hours thinking about that April morning.

"I hunted there as a kid with my father and I resent that I can't go back there and feel relaxed and enjoy being there like I used to. Somehow it's just not the same."

Familiar territory

The site of the mishap is a forested hillside on the western rim of Raystown Lake, in Huntingdon County. It wasn't far from where Grenoble was hit by several of six shots fired by two hunters. She'd first glimpsed two camouflaged hunters moving across a clearing. That's when she quit calling, not wanting to attract attention, and remained motionless against the tree. Awhile later she decided to move and upon arising was fired upon by the hunters, each emptying their 3-shot semi-automatic shotguns at the movement.

Those hunters, however, went to her aid, helped her out of the woods and took her to the Huntingdon hospital. Shirley still carries many of lead pellets in her body and extremities. Add to that the emotional wounds that haven't fully healed ….. and maybe never will.

Yet, both diehard hunters insisted on returning to the woods. For them the annual hunt is more than a hobby. Indeed, it's a lifestyle. They've both devoted considerable time to promoting safe turkey hunting and served on committees and boards of the Pennsylvania Chapter of the National Wild Turkey Federation.

Safety promoter

Ironically, it was Sandrus who spearheaded a program dubbed "Look For the Beard!" which utilizes holographs pasted to hunters' box calls, guns and other gear as constant reminders of turkey hunting safety. When moved, the green-and-black holographs show views of a hunter seated against a tree and the outstretched neck of a wild turkey. Created by Frank Piper of Penn's Woods Calls, the program was promoted statewide and nationwide. Sandrus had one of the holographs pasted to the stock of his 12-gauge gobbler gun as a constant safety reminder. Not far away, on the flip side of the stock, are holes made by the copper-plated lead pellets which also struck Ron.

Minutes prior to the shooting Sandrus had been set up against a tree calling to a distant gobbler. On the same tree he'd stretched a length of orange ribbon to declare his presence to any hunter who might be stalking his calling site. The "tie an orange ribbon" project, too, Sandrus had helped promote through the Game Commission and the state turkey federation.

"I was trying to get around the turkey so that I could call him from a different angle," Sandrus remembers." I wasn't stalking it or anything, just moving through open woods. I'd just stopped to take off my heavy sweater because it was hot. I had my gun propped against a tree and was kneeling down when it happened."

While fewer than two dozen pellets penetrated his body, others were absorbed by his gun, calls and the orange bag he carried in his game pocket used to carry a dead gobbler from the woods.

Panic sets in

"I yelled at the guy when I caught a glimpse of him going over the ridge but he never answered or came back," he recalls, still shaking his head in disbelief. "That's when panic almost set in."

Knowing he was alone and not certain how badly injured he was, Sandrus

Sandrus' shotgun shows the tight pattern of lead shot fired his way.

checked his leg for the first time and "saw the blood rushing out of the holes." He stuffed tissue paper in his pant leg and boot to soak up the blood, then worked his way back to the trail and his vehicle, parked more than a mile away.

"About half way there I checked my leg and saw it was still bleeding badly," he remembers." I sat down and that's when this strange feeling came over me. I can't describe it but it was frightening. I hope I never have to feel that way again."

But he knew he had to keep moving. Upon reaching the truck he had the wherewithal to jot down the license plate numbers of several cars and trucks parked near a power line leading to the shooting site. From there he headed to the PGC office in Huntingdon where his friend, conservation officer Wes Bower, treated the wounds until he could get to his family doctor. Game protectors in the region were quickly notified and sent to the area, but to no avail.

The intense burning sensation eventually ebbed, yielding to the sort of pain expected from being struck with shot. As with Grenoble, only a few pellets could be removed. The rest may eventually work to the surface or remain in the body for life.

The following Saturday Sandrus returned to the area with his son and a friend to hunt. He now believes that the hunter "had turkey sounds imprinted on his mind and when he looked over the ridge and saw movement figured it was a turkey. That happens in a lot of shooting accidents."

Fortunately, part of Ron's body was behind a tree and none of the pellets penetrated his neck or face. He believes the movement of his white hand (he'd removed his camouflage gloves) may have triggered the reaction of the hunter who thought he saw a turkey's head."

"I felt very depressed for a few days after that, especially about the safety program that means so much to me," Ron revealed. "I really felt let down that we're not getting the message across but I now realize that this was only one instance and it's still a damn good program."

After the season he returned to the Raystown forest with his son and found the tree where he was kneeling when the shooting occurred, then put together a scenario of what had happened. He went back a few days later – alone.

"I felt I had to go back there alone and work things out," Ron concluded. "This is where I hunted with my father more than 40 years ago."

"For my own piece of mind I really needed go back."

Hunt As If Your Life Depends On it

There's a continuing fear among turkey hunters that camouflaging yourself, then sitting against a tree and trying to sound like a lovesick hen could be dangerous.

Makes sense. Most hunters want to be seen. Turkey hunters don't. And therein lies the problem.

One Pennsylvania hunter who was shot a couple years ago "in mistake for a turkey" was advised by his lawyer that taking the case to court would be futile.

In some Wildlife Management Units, an orange caution ribbon must be hung within 15 feet of the shooter.

He was told a judge or jury would likely determine that he was inviting disaster by not providing for his own safety. Hiding himself (in camouflage) and trying to sound like a turkey with other hunters in the woods was asking for trouble, the attorney surmised.

That puts full responsibility on Pennsylvania turkey hunters to educate themselves and those with whom they hunt. Always bear in mind that "blending into the woodwork" – so to speak – and sounding like a turkey has inherent dangers.

Hunter safety and the wearing of safety orange have always been issues of disagreement between the Pennsylvania Chapter of the NWTF and the

In 2004 the PGC ruled that the fall hunter must wear 250 square inches of orange while moving from one set-up site to another.

Game Commission. How much orange is enough, too much or too little has been at the core of the controversy. The state chapter sees the solution as a continuing emphasis on education, not more mandatory orange. Some game commission personnel and commissioners believe otherwise.

Despite the arguments, both sides must be complimented on remaining civil and respectful to one another when disagreements ensue and being able to offer a handshake at meeting's end.

Safety Orange Changes

The most recent controversy was a proposal to increase the use of hunter orange by turkey hunters. However, at the October 2003 meeting in Harrisburg, testimony from the Pennsylvania turkey federation chapter and other sportsmen's groups influenced the outcome of a vote on the orange issue.

The proposal would have increased situations in which fluorescent orange had to be used. To the satisfaction of both factions, the following changes were adopted, beginning in the spring of 2004.

During the fall and spring seasons in Wildlife Management Units 1A, 1B, 2A, 2B, 5B, 5C and 5D, a hat showing a minimum of 100 square inches must be worn but may be removed when the hunter is in a stationary calling position. In all other WMUs fall hunters must wear 250 square inches of the safety material visible for 360 degrees when relocating or otherwise on the move. The orange may be removed only when in a stationary or "set-up" position. However, in these units only, 100 square inches of orange must be displayed within 15 feet of the stand or calling position.

According to Pennsylvania chapter president Carl Mowry of Butler, the issue drew more comment from hunters than any other in recent history.

"When regulation proposals are made, hunters and others have the opportunity to review them and express their thoughts to commissioners," said Mowry. "This was a great example of how sportsmen's groups can work together."

Hunting Safe and Sane

Hunter education is more than attending classes to qualify for licenses. It also calls for those of us with hunting lifestyles – newcomer or old pro – to share thoughts on defensive tactics whenever possible.

Consider these suggestions for safe and sane turkey hunting:

• Never wear any clothing items of red, white, blue or black – the colors of a mature tom turkey. Socks, undershirts, collars, boots and other apparel should all be of an earth-color or camouflage.

• Wear a head net allowing full vision; not one that binds or otherwise interferes with your eyes.

• Call to turkeys in open terrain where visibility in all directions is at least 50 yards. That way no one can sneak up on you undetected.

• Remember that not all hunters can distinguish between your calls and that of a real bird or know the difference (other than a resounding gobble-obble-obble) between hen and gobbler vocabulary.

• Use extreme caution with gobble-type calls, which may attract other hunters.

• When calling, sit against a tree large enough to hide your entire body. That way an errant movement won't as readily attract attention, or a shot, from someone behind you.

• If you see another hunter approaching, don't wave, stand or move. Instead, whistle or shout to reveal your position.

• When moving on a gobbler for a closer set-up, don't call until you're in or near a set-up site.

Always wear a comfortably fitting head net allowing full vision.

- Avoid areas known to have high hunting pressure. If you're on private land, don't presume that no one else is there.
- Never crowd another hunter working a gobbler. Turn around and walk the opposite direction when you see someone or hear his/her calls.
- When hunting with a friend, set a plan before separating so that each of you knows the general area being covered by the other.
- Never try to ambush a gobbler. Not only are chances slim of getting within shotgun range but you're inviting trouble crawling through the spring woods in camouflage.
- Use binoculars to identify far-off movements through the trees or the sources of turkey sounds in the distance.
- Mark your set-up area with an orange band (at least 100 square inches of visibility) wrapped around the tree, drape your hat or vest atop a nearby shrub or hang it from a limb. It's the law in some Wildlife Management Units.
- Look for the beard. Don't assume that a turkey moving toward you has to be a gobbler. Hens will also respond. Identifying the beard before shooting is a double-defense that assures (1) it's a turkey and (2) it's a gobbler and not a person. In fall hunting, of course, hens are legal but they still don't resemble human beings.
- Kids aren't the only people subject to peer pressure. Don't let "gobbler

fever" take command. The most dangerous hunter in the woods is the one who MUST get a bird.

• When "talking turkey" with your hunting friends, make it a point to bring up the subject of safety. You'll educate each other in the process while keeping safety foremost in one another's mind.

• Finally, don't presume that because you're an experienced hunter you're immune to being shot or shooting someone else. PGC reports show more accidents involving older "experienced" hunters than novices.

• The Pennsylvania Chapter of the NWTF also advises informing hunting partners of any health problems you may have. Don't feel you have to keep up with a guide or buddy if you're simply not in shape to run or climb steep hills. Also, inform a friend or family member of where you'll be if you choose to hunt alone.

Spring and Fall Mishaps

For the record, statistics available as of this writing show a vast improvement in turkey hunting safety in the past 10 years, with the exception of 2001 during which 53 incidents occurred.

According to Keith Snyder, bureau chief for the PGC's Hunter Education Division, the spring and fall 2003 seasons were marred by 11 shooting incidents (no fatalities); a record low nine of them in spring. Investigations showed the main reason for the accidents as "failure to properly identify the target," according to Snyder.

In the 2003 fall season only two non-fatal accidents occurred, both involving shotguns and caused by misidentifying the target.

Look for the beard before squeezing the trigger.

Although the goal is not to have any mishaps afield, the 2003 figures remain encouraging, especially when compared to 2001's 53 accidents. In 2002, 24 shootings – including one fatality and one self-inflicted wound – were reported.

The second factor influencing the accident rates was "in line of fire," mishaps, Snyder's records reveal. Each involved shotguns. No fall season rifle incidents were reported.

We all think we're safe hunters. But consider that most of the victims, and the hunters who fired the shots, surely went afield with the same attitude: "It can't happen to me."

Think again.

It can.

TIM FLANIGAN
Bedford, PA

Tim Flanigan has been a wildlife
conservation officer with the PGC for
30 years. An award-winning wildlife
photographer, he's given up carrying a
shotgun when he guides friends on gobbler
quests in Bedford and Somerset counties.

TURKEY HUNTING'S SINS
ALL SPELL DANGER

I enjoy a unique perspective with regard to game law violations because I
am first and foremost a sportsman, but I also happen to work as a Wildlife
Conservation Officer (WCO) in Bedford County.

Prior to entering the profession I had hunted for many years and had become
a proficient turkey caller. I have been nearly shot while hunting turkeys and
as a WCO I've investigated numerous shootings of turkey hunters. Due to
those experiences, I believe that the most severe violations are those concerning
safety. Although I, and most sportsmen, deplore violations of fair chase such as
hunting from illegal blinds and shooting birds on bait, and condemn additional
turkey losses due to illegal killing, the maiming of a human or the loss of a
human life renders safety regulations the most important of all.

No Orange: Most Common Violation

Turkey hunters have been schooled to believe that wild turkeys are virtually
impossible to kill due to their exceptionally acute senses, especially eyesight. Because
of this, the most common turkey hunting violation is the failure to wear protective
orange clothing.

During the spring 2004 gobbler season I did not observe a single hunter obeying
the orange regulations while they were afoot in the forest. It is amusing to observe
them don the required orange garb as they near the highway where their car is parked.
Many of these hunters had no orange clothing in their possession or even in their
vehicles. Such is the degree of disdain for this regulation – a simple regulation that has
prevented more accidents and saved more lives that we can ever know.

I observed one pair of hunters move about the forest for more than an hour
while wearing full camouflage. One of the men wore a black hat covered with a
black head net. This was the first thing that I spotted and I initially thought that
his head was a turkey at 200 yards. While observing them, I was horrified to see
one man pump five shells out of his gun while its muzzle was pointed at the other's

chest. Near the highway, they put on their orange vests and hats and claimed to have been wearing them all morning. This is so very typical.

The Danger of Stalking

The regulation that is not fully understood and very commonly violated, often with disastrous consequences, is that hunting is by calling only during the spring season. Stalking is illegal. You are never alone in the forest and only a fool will let himself believe that he is. All too often hunters seek to stalk within gun range of turkeys while calling, then shoot at the first

The most common turkey hunting violation is not wearing orange while moving.

movement seen. These hunters are often inexperienced turkey hunters who will stalk either hen yelps coming from a real hen or a caller set up nearby. For this reason I never carry or use a gobbler call in the forest.

Other Scary Violations

Other common violations include unplugged shotguns and the use of illegal shot sizes. In one hunting accident, the victim and offender were stalking each other when the victim was shot in the face, neck and hands with one No. 6 shot barely missing his carotid artery. It is fortunate that he was the victim and not the shooter, for his 10-gauge held hand-loaded shells containing .22 caliber ball bearings. This was a fall season hunt.

Hunting turkeys over bait is common and seems to be increasing. Failure to tag and attempting to take over the limit are also common violations, as is roost shooting.

These are greedy hunters and they are the most dangerous because they will take all types of chances to simply kill. The essence of the hunt is not even a remote consideration for this type of hunter. This hunter can kill you and this is why: We need to realize that we see with our minds – not our eyes. What we see is in actuality the interpretation of the light gathered by our eyes. The eyes are simply light catchers. That is how we tell a Corvette from a Mustang or a hen pheasant from a cockbird. If we wish, hope, desire and obsess about bagging a turkey, our minds will see shadows and movements as big gobblers with long beards. A few added turkey sounds will trip his trigger. Only a fool will say, "That can't happen to me."

It's the Experience First

All turkey hunters need to assess their ultimate goals prior to entering the woods. If that goal is to enjoy the experience to its fullest and come home safely, the hunter will enjoy many days and years afield and learn more about turkeys on every trip.

CHAPTER 16

PATTERNING YOUR SHOTGUN
Know Where Your Gun Shoots

*Patterning a shotgun requires taking some shots with a bench rest
and others with your back against an object.*

Before you head to the woods, know where your shotgun shoots.

It has always baffled me as to why some hunters will spend hours sighting in and shooting their deer rifles before the buck season opener but will ignore this most-important duty before the turkey seasons.

Deer hunters sighting in new scopes will pick up a box of shells, go to a range and adjust their scopes accordingly, then practice with a few more shots. Many turkey hunters ignore this basic duty because: (1) They don't think it's necessary to "sight in" a shotgun; (2) They believe the cone-shaped spread of a shot load as it travels from the muzzle will be sufficient to fatally hit a turkey at 30 yards; (3)They'll take a couple offhand shots at a turkey target 20 yards away and if some of the pellets hit the center, believe they're ready to go; and (4) Whatever turkey load they may have picked up on sale at the local Save Mart will do the trick.

Then they can't figure out why they missed that longbeard at 25 paces.

So much for the lecture.

How to Pattern Your Shotgun

Patterning a shotgun is a relatively easy – and enjoyable – task. The exception is the stubborn gun that doesn't throw a satisfactory pattern with any of the shotshells you may be trying out. Should this be the case, a trip to the gunsmith may be in order.

Kevin Howard of Winchester checks out his 30 yard pattern prior to heading afield.

Here's the routine:

• Measure 30 yards and place a target simulating the actual size head, neck and upper breast of a turkey. The targets are available at most sporting goods stores or you can make your own. Place the target in a site with a safe background. If your turkey target shows only the head and neck, tape it to a 30x30 inch (or more) sheet of paper or cardboard. This will enable the shooter to see where nearly all the pellets hit, including those that struck the body or missed the head target altogether.

• To reduce human error (such as flinching), stabilize the shotgun and aim at the spot where the feathered breast joins the bare neck.

• Repeat the procedure on a second target.

• Count the holes in the head and neck area on each target. If you know the number of pellets in the shell, count the punctures in a 30x30 inch background to determine the approximate percentage of shot which has at least some potential for striking vital areas. It's generally agreed that a minimal 6-8 pellets must penetrate a turkey's vital area, more with smaller shot sizes.

Birchwood Casey's Shoot-N-C targets highlight the shot holes.

• If the shot pattern is clearly skewed, try a second or third test with different shotshells.

• Once a satisfactory load has been established at 30 yards, take two or three offhand shots each at 20 and 40 yards. Also do a test with shots at 10-15 yard targets to compare the dramatic pattern variations when shooting at close quarters. You may be surprised, as I was, as to the density of a full-choke pattern at 10 yards.

• Finish the project by simulating an actual hunting set-up, taking aim with your back pressed against a tree and your gun braced on a knee. Now you're ready to hit the turkey woods with the necessary knowledge and confidence that the pellets will pattern where your shotgun's aimed.

115

DAVE STREB
Cuba, NY

Dave Streb is the v.p. of sales and marketing for Quaker Boy Game Calls. He's hunted turkeys in 27 states, has six Grand Slams to his credit and gives turkey seminars throughout the nation. Back home, Dave hunts McKean and Potter counties.

JUDGING DISTANCE CRUCIAL

You did everything right. He just gobbled again, closer this time. Your heart's pounding. Now you catch movement. You see he's a longbeard. Your heart shifts into high gear.

You tell yourself to let him come. Then he stops and stretches his neck. Does he see me?

You shoot and off he goes.

A rare opportunity and nothing to show for it. You pace off the distance and are surprised it is about 48 yards.

You're upset because your "turkey gun" and "turkey loads" are exactly what your buddy uses. He always talks about killing birds at 50 plus yards. What went wrong?

No two guns pattern exactly the same. It's up to you to determine the maximum effective range of your shotgun. Try a variety of shells and shot sizes at varying distances. When you get at least 10 pellets in the head and neck, pace off that distance. That's the maximum effective range of your gun.

The optimum range is also important. At what distance is your entire pattern inside a 15 inch circle? It takes some serious shooting to find out.

Now that you know these distances, how do you judge them as you walk in the woods?

The off-season is a good time to practice distance judging. Pick out a tree, rock or other object and take a guess how far away it is. Then pace it off with a standard step length. Always travel in a straight line and remember that judging distances in a field and in a woods may be deceiving.

How good are you at this skill?

If you're not plus or minus two yards every time, I suggest investing in a range finder. Pick a spot that marks your maximum effective range and another at your optimum effective range.

If you're using decoys, place your closest one (preferably a jake) at your optimum effective range and the furthest one at your maximum effective range.

Remember, when that turkey stops and puts his head up, it doesn't necessarily mean that he's alarmed. Most times he's just looking for the "hen" ..... you.

TERRY HYDE
Tyrone, PA

Terry Hyde has used his Ithaca Model 37 16-gauge for the past 37 years. He's taken gobblers with it in Blair, Huntingdon, Centre and Cambria counties and knows its precise shot pattern. Terry is a recipient of the Roger Latham Award from the P-NWTF.

DEDICATED TO HIS SWEET 16

In the modern world of turkey hunting, hunters have been molded into a system of specialty chokes for special loads coming out the barrels of 12 and 10 gauge guns to overmatch any turkey from 40-50 yards.

I applaud modern technology, but I've never fit into that mold. For 37 years I've carried my trusty Ithaca Model 37, a 16 gauge shooting only 2 ¾ inch shells when after turkeys. And I know precisely how it patterns and its limitations.

I did do some modifications after hunting with it for nearly a dozen years. I sent the entire gun back to the Ithaca factory to have a full-choked barrel fitted. This was prior to the advent of screw-in-chokes. It replaced my old modified barrel, which prior to adding the choke claimed several gobblers for me.

I also switched from the standard 16-gauge lead magnums to a special No. 6 copper, hand-loaded magnum from a gentleman in Maine (copper plated buffer shells were not made for 16 gauges). It complemented my tighter and longer pattern for turkey hunting.

That has been my rig for over 30 years and it suits me just fine. It has a terrific pattern out to 30-35 yards but I prefer turkeys in real tight, like 15-20 yards. Maybe that's because in my days with the modified barrel I knew I had little killing pattern after 20 yards.

The first spring gobbler I ever shot was at 16 steps. I knew I had to get him in close so I just kept calling. Therefore, my tip to you is; once you see the bird coming, keep calling if necessary. Too many hunters stop calling the instant they see feathers and the bird hangs up. Calling after the bird hangs up just seems to make the gobbler all the more wary.

I prefer to tone down my calls while keeping him enthused. He'll come "right up the barrel." I have killed many turkeys inside 15 yards by just continuing to call.

And believe me, a loud throaty gobble at 10 yards sends the heart pounding. It's an eye popping experience that I live for every year.

So will you the first time one seems ready to climb on your lap, whether you're aiming with a 16 gauge or not.

CHAPTER 17

WHY WE MISS

It Comes Down
to the Moment of Truth

It's the moment of truth.

A gobbler has answered your yelps and cutts and he's on his way toward you. You suspect it's the same bird you watched fly down from his roost an hour earlier. You haven't seen or heard any hens lately and you suspect he's coming back – cautiously and silently.

You adjust your mask, settle the gun across your knee and squint across the double-bead sights of your pump gun for a couple seconds, then try to relax,

It's the moment of truth. Are you up to making the shot?

which you find yourself able to until a bobbing white head is spotted through the brush. He's 60 yards out and he's only gobbled once. You offer another series of subdued yells and he finally responds, sending a tingle down your spine and urging a choking grip on your gun.

You pick out the spot 10 yards behind your decoys in which he's likely to appear, click off the safety ..... and await his appearance as he slips behind a deadfall, only his bobbing, white head showing his location. Seconds later he appears at 25 yards, seemingly standing on tip-toes as he eyes the artificial jake and hen. You slowly shift your gun a few inches, squint down the barrel, set the bead on his neck and squeeze the trigger.

In a second he's airborne, climbing swiftly over the leafy canopy and disappearing into the greenery. You sprint to the place he was posing seconds before to check for blood or feathers, but you know what happened. In the captivating seconds leading up to the moment of truth, you concentrated too much on the tom's colorful head and fan-dance and forgot about properly cheeking your gun stock and laying the front bead flat on the barrel. Indeed, the bead was on the bird but the muzzle was tilted upward, sending your load of No. 5s over the longbeard's head.

You blew it and now have the same knot in your gut that's been experienced by turkey hunters since enough turkeys were around to be hunted.

Your next step is to give some pre-hunt thought to this depressing matter so it won't happen again.

First off, chalk it up to experience. I eye with suspicion the guy who boasts that he's never missed a turkey. It's happened to all of us and it doesn't take long for that hang-your-head feeling to disappear. By tomorrow you may even be able to laugh at it as you tell the crew at work what occurred.

Then again, perhaps not.

How and Why We Miss

The fictional scenario above isn't the only way to miss a tom, although shooting over its head is arguably the most common error.

Jason Morrow has done his share of videoing many of those misses and saw what occurred more than once.

"The number one reason turkey hunters miss is that they get excited," said Morrow, producer of the NWTF's Get in the Game television show airing on The Outdoor Channel. "It's an exciting hunt, no question about it, but calmly and carefully going through the routine of pulling that trigger comes with practice."

One big problem is that some hunters want a better look at the bird as they pull the trigger.

"It's pretty common to see a hunter pull his or her head from the stock of the gun to get a better look at the bird when they shoot," Morrow said. "Once they pull their head, their eye is not straight down the barrel and more than likely they'll shoot above the bird."

The All-Important Set-Up

Shooting above or to the left or right of a turkey is but one gremlin causing errant shots. For instance, setting up in a comfortable position for a shot is one of the most important elements when turkey hunting. It's more than just sitting against a tree and hoping the bird will show up in front of you.

"If you're right handed, you want to set up so the bird would most often approach from your left or straight ahead," Morrow advises. "You don't want to make too much movement because of the turkey's keen eyesight, so you

Right-handed shooters should set up so the anticipated approach of the bird is slightly to the left, permitting a 180-degree swing. For lefties it's just the opposite. Remaining calm and resting your gun on a knee are key to letting a gobbler work into range.

want to be set up so your shot is a comfortable one. Sometimes it can be a long wait and comfort is the key to you staying still."

Setting up as Morrow recommends will give you a full 180 degree swing. Try shooting to your right, however, and you'll be more limited.

Remember also that if you need to move your gun and the turkey has the potential for seeing you do so, make certain that his head is hidden by a tree or some other obstacle. Then make your move. A turkey has a phenomenal 300-degree field of vision and even though he or she appears to be looking at something else, it can still see a quick movement.

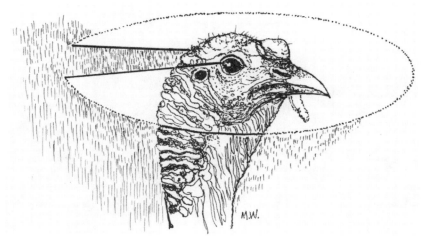

A turkey has a phenomenal 300-degree range of vision.

Tips for Avoiding Misses

Without question, missing a turkey is the number one "downer" in the turkey woods, no matter if you're alone or with a friend. Here's some helpful advice from the NWTF to help avoid a miss in the first place.

• Practice shooting from the same positions you would in a hunting situation. Wear your pocket-filled turkey vest, head net, gloves and any other mandatory gear to make sure nothing catches when you shoulder the firearm.

• Pattern your shotgun from different distances. Each gun, choke and turkey load combination has its own pattern and range. Eight to 10 pellets in the vitals at 40 yards should be enough to kill a turkey.

• Ultra-tight choke tubes are not necessarily better, depending on the circumstance. While the new super-choke guns offer high velocity and knock down power at greater distances, reconsider things when a bird gets in too close. Tighter choke tubes can constrict the shot pattern so much that the pellets will not have enough time to expand if a bird is too close to your setup. Close shots with ultra tight chokes may be the mere size of an orange or grapefruit.

• When you have time to set up, choose a tree that is wider than your back and taller than your head. Make the same considerations when practicing under mock hunt conditions. Not only is this the safest setup, but the tree will break up your outline and provide a solid back rest when you take the shot.

• When its time to shoot, wait for the gobbler to come out of full strut and extend his neck. When a tom's strutting, his spinal column is compressed and his head is tucked and partially hidden by his feathers. This makes for a smaller target area.

Most of this is common sense, but when that 20-pound longbeard with its crimson or snowball head aglow pops into shooting range, we don't always perform as we planned.

Blame it on turkey fever ….. and fight it.

DON HECKMAN
Camp Hill, PA

Don Heckman is well known in NWTF circles and has devoted
countless hours of service over the years. He's also hunted turkeys for
more than three decades and has pursued them in 13 Pennsylvania
counties. He has taken the Royal Slam of five subspecies in the U.S.

Preparation the Key to Avoiding Bad Shots

Missed turkey.

Wounded turkey.

They're words that no hunter wants to hear, but it happens to all of us sooner or later. Factors leading to those words often start with the shooter and the firearm or bow in an unfamiliar position. For example, if a turkey hunter is not in a comfortable sitting position or is caught in a position where he or she has to twist or turn in order to get a shot, it's more likely to result in a miss or a poor hit.

Woundings and outright misses are more likely to happen to hunters who do not regularly pattern their shotguns or practice properly with bow and arrows. Shooting 3-D deer targets with archery gear will undoubtedly hone your abilities but aiming on a hen or gobbler is a different matter. If you're serious about bowhunting for turkeys, a 3-D turkey target will be a wise investment.

Also, patterning your shotgun at different distances and with different loads is a must. Take aim by shooting from different set-ups and shooting positions (sitting, kneeling and standing). At least once, wear the same clothing you'll be wearing afield.

On actual hunts, taking wild shots or shooting through thick cover are surely not recommended with bow or gun. Always be aware of your shooting lane and what is beyond your target. If a killing shot does not present itself, do not take a chance on simply "letting one fly." Instead, hold your shot as long as possible and await the right moment. This should be practiced on the range so when it occurs in an actual hunt you'll know how long you can hold a draw.

Shotgun hunters may be tempted but should not take body shots. Head and neck only is the rule. For bowhunters, the head/neck shot is a tough killing shot and should not be attempted. The slightest head/neck movement, or turn, and you might wound or totally miss the turkey. The latter, of course, is preferable. The targets for bowhunters are broadside through the lower portion of the wing, head-on into the breast or, if the bird's facing away from you, between the shoulder blades. They are all acceptable.

In the event you wound a turkey, quickly survey the situation. If the turkey is flapping its wings wildly but cannot regain its feet, you should have an excellent chance of getting to it and retrieving the bird before it runs or tries to take flight. A body-shot turkey usually rolls and flaps its wings, and may run or attempt to fly away. A quick second shot, or even a third shot, may be needed to kill the wounded turkey. If it happens to gain flight (which it can do in an instant) and disappears in the treetops as it glides down a slope, chances are you will not find it. But the ethical hunter will look for it anyway.

CHAPTER 18

IT'S YOUR CALL:
A How-To Guide to Buying Turkey Calls

I hate shopping.

In fact, I despise shopping. Being confined to a mall is as harsh a punishment as can be imposed on any red-blooded American man, especially one who'd rather be outdoors doing something useful like hunting or fishing.

Of course, if the mall has a hunting and fishing store I have been known to endure an hour or so of comparison shopping for a gun, boots, a new binocular or maybe another two or three additions to my turkey call collection, which already fills several large drawers in my basement "escape room."

The first turkey call I ever bought was a Lynch box call that served me well. I still have it, as a matter of fact, but it's relegated to the top shelf in my trophy room for nostalgic reasons even though it still performs quite well.

Well-known turkey hunter and callmaker Louie Stevenson of Wellsboro holds the gobbler he shot in 1978 using his Turkey Talk box call.

That was followed with a slate-and-corn-cob peg device given to me by longtime friend Bob Clark of Mechanicsburg more than 25 years ago. Several seasons later I picked up a diaphragm mouth call which, on my first outing, I nearly swallowed when I slid down a slope and sucked in some air. After gagging for a minute or two, which seemed like an hour, I plopped myself chest-down atop a deadfall three or four times, finally managing to free it from my throat. The thought of needing a Heimlich Maneuver and being alone still rings clear in my memory. Squirrels don't do Heimlich's ...

It wasn't until about a dozen years ago that, at the encouragement of a master caller, I developed the confidence to place a diaphragm back into my mouth. Since then I've been able to again produce adequate hen music with the horseshoe-shaped instrument, although I keep it in my pocket when in company of some of the turkey calling pros I meet at camps and calling contests.

On that initial shopping trip for a turkey call 30-plus years ago my choice was limited to about four or five models – all either box or slate calls. I also used a self-fashioned snuff can call made, instead, from a piece of latex stretched across a 35mm film aluminum container.

Today well-stocked stores and the myriad of sporting catalogs showing up at my door present a wide and often confusing array of instruments, all promising to bring a boss gobbler to 25 steps.

Hunters once fashioned their own "snuff can" calls or made them from aluminum film canisters and latex

As with gun and garb choices, picking the "right stuff" for serenading lusty gobblers is largely an individual matter. Wise shoppers must go armed with some basic knowledge before putting down the cash for a call of the wild.

Calls can be divided into two broad categories: calls that produce sound by friction and those which operate by air flow.

Here's a shopping guide of pros and cons on the different varieties of calls and an overview what's available in today's turkey call marketplace.

Mouth Diaphragm Calls

Description: Small, horseshoe-shaped aluminum frame covered with thin plastic. Frame holds one to four latex reeds which vibrate when air is blown under them. Calls are held against the roof of the mouth with the tongue.

Advantages: Small and easy to carry. Inexpensive. Wide variety of single and multi-reed calls available. Produce realistic sounds. Easy volume control. Hand movement kept to an absolute minimum. Can be directional with use of cupped hand against the mouth.

Disadvantages: Among more difficult of calls to master. Requires constant practice. May not fit all mouths (due to dental work or odd palate configurations).

Causes some users to gag. Remote danger of swallowing. Easily damaged by improper handling and storage.

Commentary:

"The biggest problem with diaphragm calls is that everybody tries to make it too hard to use them," says Will

The diaphragm call is the modern version of the old songbird call first made in the late 1800s.

Primos, whose Primos Wild Game Call Company produces a variety of mouth calls.

Primos is the mentor who sat me down, showed me how to overcome my gagging reflex, and had me clucking and yelping within minutes. Part of the lesson was demonstrating how to trim a call for a proper fit on my palate. He recommends that newcomers to calling first purchase diaphragm calls on which the frame is sufficiently large to allow customizing. He also suggests starting with a multi-reed device.

"The single reed call is the hardest to use," said Primos. "My recommendation is to start with a double reed or a stacked frame call."

Will Primos recommends the multi-reed call to newcomers wishing to learn mouth calling.

Calls with 2, 3 or even 4 reeds bring intensity and raspiness to the sound. The layered reeds are staggered or "stacked" one behind the other, enabling a wide variation in the sounds they produce. Some also have notched reeds (called split-reed calls) for variations in tone and raspiness.

As the name implies, stacked frame calls consist of two frames, each holding latex reeds. The thin reeds vibrate independently, offering a more realistic quality. Primos considers them the easiest to use and the best for the novice caller.

While many hunters are intimated by the mouth-call "magicians" they hear at calling contests, Primos encourages every hunter to learn how to use a diaphragm call. He believes "the easiest person to teach is someone who has never used one before."

Will recommends first learning the "cluck" – the easiest call to mimic and an effective single-note gobbler enticer.

"Do it with your lips," he instructs. "Put the call in your mouth with your tongue against it, then "spit" the letter P by huffing air from your own diaphragm."

Calls which can be reproduced on a diaphragm include yelps, cutts, purrs, clucks, cackles, whistles and kee-kee runs, the latter an effective fall call. Some skilled callers can also imitate an acceptable (at long ranges) gobble.

While the diaphragm call is thought to be a modern device, former Turkey Call magazine editor Gene Smith says the basic tool (a bird call) was made in 1867 and the initial turkey diaphragm was first marketed in 1923.

Tube Calls

Description: Basically a diaphragm call held outside the mouth. Hard plastic, flared tube with oversize latex reed stretched across bottom half of one end. Latex position and pressure adjusted with rubber band or removable rim. Air is "huffed" through tube while placing lower lip atop the latex.

Advantages: Can be used by hunters who have difficulty "mouthing" diaphragm calls. Capable of reproducing all turkey sounds, including the gobble. Ease in rendering ultra-loud and soft sounds as well as high volume on windy days.

Disadvantages: Requires at least one hand (and therefore some movement) to operate. Exposed reed easily damaged (most calls come with extra reeds). Not as widely available as most other calls.

The tube call is often used by hunters who do not feel comfortable with the diaphragm call.

Commentary: Although many hunters consider it a new addition to the marketplace, the tube call is actually one of the oldest of calls.

"It's really a modern variation of the old snuff can call," said Dave Hale, half of the amiable Knight & Hale Game Call Company team. "I'd say it may be the most versatile of all kinds of calls in that it can make every known turkey sound."

Hale considers the tube call "a musical instrument" that requires a certain amount of practice to master – more practice than most other calls, save for the diaphragm.

"For some reason this is one of the least popular calls," said Hale. "But don't tell that to people who have learned how to use it."

A tube call is capable of reproducing the best gobble of any type of call – so good that from a distance and against the lips of a talented caller it's sometimes hard to distinguish from the real thing.

Hale likes tube calls for their ability to "reach out." On windy days or when trying to get a rise from a distant bird, the tube call has no equal. He's also

Realtree's Bill Jordan uses the tube call for loud, long distance calling.

discovered that turkeys who hear many calls during a single springtime are less call-shy when they're beckoned with tube calls. The latex reed is easily adjusted to alter the pitch and raspiness.

"This is definitely an under-rated call," Hale concludes. "I'm not sure why it's not more popular but that may change as more hunters give it a try."

Hinged Lid Box Call

Description: One of the oldest types of turkey calls. Sound produced by beveled wooden lid scraped across top edge of box. Built of various woods, usually cedar, walnut, mahogany, cherry or poplar. Tone variations produced by different species of woods. Higher mellow sounds offered by hardwoods. Softwoods render deeper sounds.

Advantages: Among the easiest of calls to master. Produces all sounds except fall "whistle" or kee-kee. Long life span. Wide variety on the market.

Disadvantages: Large and sometimes cumbersome. Noisy when carried improperly. May be affected by

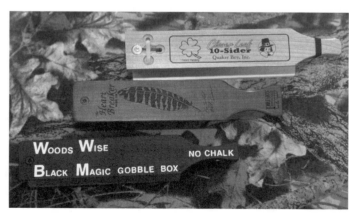

The hinged lid box call is arguably available in more styles than any other turkey call.

damp weather. Will not function when wet. Requires use of both hands. Hence, movement by hunter.

"If the only call you owned was a box call you could do it all," says Dick Kirby, "boss gobbler" at Quaker Boy Game Calls in western New York.

For the box call shopper, Kirby recommends a request which may not please some sporting goods store clerks.

"Ask to take it out of the box and try it," he advises. "That's the only way you'll know if you like it. I encourage people to do that with my calls."

Simply knowing the species of wood with which a box call is made provides little information as to its quality. Many calls, Kirby explains, are made with two woods. One Quaker Boy box holds a cherry paddle (lid) and poplar sidewalls. Another is fitted with a cedar frame and maple paddle. Pitch, raspiness and volume are all affected by the wood type.

Dick Kirby uses his Quaker Boy Boat Paddle Call on a windy Texas gobbler hunt.

The shape or arc of the paddle's underside, which is scraped along the top edges of the box, influence what Kirby calls the "breaks" in a particular call; such as the two-pitched "yee-elp" sound. Pressure on the paddle, roughness of the wood (Kirby uses a medium 100 grit sandpaper on the paddle and top edges of the box instead of chalk) and the speed of each stroke all add variety to the sounds. Most callers use soft carpenter's chalk on their boxes to increase friction.

A box's side walls determine the call's pitch: the thicker the wall, the higher the pitch. Sounds become lower on thick-walled boxes.

Kirby also cautions that gripping the box in varied ways can change its tone and volume. The standard technique is to firmly "cuddle"

This River Valley call and some others go a step beyond the traditional call with artistic renderings on the lid.

the bottom of the box with one hand while holding the paddle between the thumb and index finger with the other. Use only the wrist, not your entire arm when stroking, he advises.

Few, if any, box calls leave the factory without hand-tuning by experts with an ear for turkey sounds. Most companies will readily replace any call deemed faulty, according to Kirby.

"I don't know of any turkey hunter who doesn't own a box call," Dick explained. "No two turkeys sound the same and no two box calls sound the same. I can almost always tell if a sound is made by a man or turkey but I have been fooled by box calls."

Push-Button Box Calls

Description: Palm-size box call without lid. Pyramidal striking surface glued to inside bottom of box. In-line striker attached to push-peg and positioned against sound-producing surface. Pressure on peg and striking

The push-button box call is the easiest to learn.

surface controlled by piano-wire spring, coiled spring or rubber bands. Both friction surfaces require chalking.

Advantages: Considered the easiest of calls to use. Excellent choice for beginners. Requires only one hand to operate. Can produce yelps, clucks and purrs. Can be fastened to barrel with plunger moved by a length of twine for use with firearm in shooting position.

Disadvantages: Difficulty in keeping spring "tuned" or in place on some calls. Minimal variation in volume and tone. Ineffective when wet.

"This is a sure-fire call for giving a new hunter confidence," says Terry Rohm of Wellington Outdoors. "I even see experienced hunters using it now."

Rohm believes the ease of use of the push-button call was a deterrent to its acceptance by callers when it first appeared. It was simply too easy to use – but very

The push-button call is also sold with barrel-mounting clips and activated by twitching a length of twine.

effective for novices and others who had not mastered more complex devices. Once its versatility was realized, however, more members of the turkey hunting fraternity adopted it as part of their arsenals.

"It has a different sound than a box call but is quite realistic," Rohm explained. "I don't know of any other call on which you can get the rhythm down so easily." Sounds are made by simply pushing and applying pressure to the wooden peg. A spring or rubber band supplies resistance which quickly returns the peg to its "trigger" position. Hen yelps, purrs, cackles and clucks can all be imitated. By holding the palm of a hand over the box or using a forefinger to apply pressure to the striker, the pitch, tone and volume can be varied.

Several years back the introduction of the "Fight 'n Purr" from Knight & Hale kicked off a variation on the push-button theme. The double-striker box call imitates turkeys quarreling and will bring in curious gobblers enticed by the sound. It's a great call for hung-up gobblers.

"The push-button call requires very little movement and that's a big advantage when a bird is within your view," said Rohm. "I also recommend it as a kid's first call."

While the push-button call was once considered a toy, today it's gained the respect of thousands of hunters who accept its limitations while praising its effective simplicity.

Slate/Glass Friction Calls

Harrisburg resident Matt Morrett of H.S. Strut has won dozens of contests using glass and slate calls.

Description: Consists of two parts: slate or Plexiglas surface and striker (also referred to as a peg). Strikers are made of plastic, acrylic (also a plastic), Plexiglas or wood. Oval or round slate or glass usually fitted into wood or plastic rim for easy handling. One variety composed of two layers of slate or glass with separation creating a resonance chamber. Surface must be roughened regularly with sandpaper, emery cloth or an abrasive pad.

Advantages: Relatively easy to use. Potential for wide variety of calls using different strokes and pressures on the peg. Varied strikers produce variety of sounds. Glass call can be used in wet weather with some exception. Long-lasting with proper care. Two hands needed for use.

Disadvantages: Prone to "strange" sounds when surface is not roughened or peg "slips." Contaminants such as oil from hands and moisture will affect slate, causing the striker to "slip." Both hands must be used to operate. Movement necessary.

Matt Morrett of Hunter's Specialties' H.S. Strut Pro Staff is a master of these friction calls, having won many contests with the glass, slate and peg devices.

"The cost of slate and glass are pretty much identical," said Morrett. "It's an individual preference but I'll give the edge to the glass calls, even though I carry both."

Morrett said the slate produces a "softer and more mellow sound" while the glass surface is "a bit higher pitched with potential for more volume" than slate.

Greg Neumann of Penn's Woods Calls gives Betty Lou Fegely a lesson in using a glass call.

He credits famed Pennsylvanian call-maker D.D. Adams as the innovator of the sound chamber in the slate call, a concept adopted by nearly all call manufacturers today.

Double-slate and double-glass calls offer the advantage of better resonance and increased volume. Single piece calls are typically more subdued although volume can be controlled on all such friction calls by lightening pressure on the striker and holding it close to the body.

Of utmost importance is the regular (several times a day while hunting) roughening of the surface. The bottom of the striker pegs should be sanded for maximum effectiveness.

A slate surface and wood striker are useless when wet although an acrylic or Plexiglas striker will draw sounds from the call. A better bet is a non-wood striker and a glass call on drizzly days.

Morrett recommends testing two or three strikers and choosing the one that's most pleasing for a particular call. He prefers pegs with flared or enlarged tops for better control and enhanced resonance.

When working the call, do not lift the striker from the surface. Use smooth, circular or straight-line strokes, depending on the type of hen sound desired. Each call has a "sweet spot" which should be located and marked (with a scratch or notch on the rim) for quick reference afield.

For maximum volume, Morrett recommends holding the call at arm's length. To produce subdued calls, arms and hands should be tucked near or against the body.

"You can do cackles, clucks, cutts and purrs on a glass or slate call," says Morrett. "It's all that many hunters use."

Wingbone and Gobble Tube

Of course the oldest turkey call (next to natural voice and stiff grass blades) is the wingbone, which Indians and pioneers adapted from two or three bones from a turkey's wing.

It is, perhaps, the most difficult to master, serving as a challenge for experienced hunters, some of whom make their own wingbones. The sounds produced are limited only by the talent of the caller, although the cluck and yelp are basic.

The small end of the 6-8 inch long call is placed between the lips. The sound is made by a sharp sucking of air emitted through the opposite end, which flares with the natural shape of the larger bone.

Several companies produce commercial wingbones (made of plastic, wood or metal) for hunters wishing to take on a traditional challenge. With a bit of searching or contacting your local P-NWTF chapter you'll be able to find someone to make you a real

Outdoor writer Jim Casada makes turkey music from his handmade wingbone.

wingbone call from your last gobbler. Many of these makers impart an artistic look to the functional calls.

Consider also a call such as Pennsylvanian Red Wolfe's Gobble Tube which, when shaken, emits a realistic gobbler imitation. The call is built into a rubber cylinder; basically a sound chamber with two diaphragms. It also makes the cluck, putt, yelp and cackle. It works, particularly when a gobbler ignores everything else.

On any shopping spree for calls, also check out the optional items such as storage containers for reed calls, a band, sleeve or holster for box calls and, of course, an audio or videotape for those practice sessions which will help you learn to master these calls of the wild.

Oh, yes. Here's one more way to get the most from your turkey calls.

Some 20 years ago during a hunt in Alabama with legendary turkey hunter Ben Rogers Lee, the subject of learning how to effectively use your arsenal of calls came up during a dinnertime conversation.

Big Ben suggested: "If you buy a call and can't learn to use it, at least get your money's worth by driving your wife and kids nuts with it every once in a while."

We've all done it.

RALPH MARTONE
New Castle, PA

Ralph Martone is a recipient of the Frank Piper Hunting Safety Award. He writes a weekly outdoor column in the New Castle News and serves as president of the Cascade Thunderin' Toms chapter. Ralph's turkey treks are focused in Venango and Lawrence counties.

SLATE OR GLASS: WHICH FRICTION CALL IS BEST?

Even though I carry glass and aluminum friction calls, I still consider the traditional slate call my "go to" call on a spring hunt. Slate calls can accurately reproduce most turkey sounds in a wide range of volumes. From the loudest locator yelps to the quietest finisher purrs, the slate call can take a gobbler hunt from start to finish.

Slate calls can be tuned to match most turkey sounds. Your choice of striker will change a call's pitch from the deep, chesty sounds of an old gobbler to the excited yelps of a young hen. In general, a slate call's pitch is determined by the striker's hardness. Softwoods like cedar, fruitwoods and chestnut make low-pitched tones, while hard woods like maple, walnut and many of the exotic woods produce higher pitched, more excited calls.

In addition, a slate call can be fine-tuned by calling from different parts of the call. Calls produced from the slate's center will be rich and ring in the air longer than those from near the edge of the call.

My all-time favorite slate call is a wormy chestnut pot with a hard maple striker made by a couple of "good old boys" from Zelienople in western Pennsylvania under the name Talkin' Stick Calls.

When the weather turns wet I turn to the glass and aluminum friction calls. These calls work well regardless of the weather. Both the glass and aluminum surfaces produce higher pitched calls than the traditional slate call, which will occasionally trigger a response from a tight-lipped gobbler.

A glass striker worked on a glass call can be a deadly combination in wet weather. In addition, carbon strikers produce excellent turkey yelps and purrs on both glass and aluminum calling surfaces. My favorite glass call is made by another Pennsylvania company, Cody Turkey Calls of Halifax.

DALE ROHM
Blain, PA

Dale Rohm is known across the East for his quality line of private label box calls. His biggest tom in 55 years afield is a 24-pound New York behemoth with a 12-3/8 inch beard and 1-5/8 inch spurs. Perry County holds his favorite hunting grounds.

QUALITY WOODEN BOX CALLS NEED SOME T.L.C.

Shopping for a quality box call?

If possible, the call should be made from a wood product. This makes the call a lot easier to tune and is more consistent in producing all the turkey sounds needed to call in a bird. Choose a call with grooves on the side of the box. This will make a vibration to help give you the double note that makes the box sound like a real bird.

The type of wood used makes a difference in the sound. Most woods will work with a good quality box call. The most popular woods are walnut, cedar, cherry, mahogany, poplar, and purple heart -- or a combination of any of these woods.

Next comes the thickness and radius of the underside of the paddle. It should be the same thickness as the radius on the top lip of the box. This is one of the most important features in building a quality box call.

The glue in the call is also a big factor in getting the desired sound. If the glue does not bond well to the call, it will sound flat or squeaky.

After investing in a box call, some good maintenance is in order to maintain its condition. If your call does not have a finish on it, put a light coat of tung oil on it. This oil is from the tung tree and gives a waterproofing finish. It will not change the sound, but be careful not to get any oil on the underside of the paddle.

Store your call in a cool dry place so it does not draw moisture and keep it out of direct sunlight for long periods, such as tossing it on the dash of your vehicle.

It's also recommended that you buy a call holder in which to store and protect it.

To change the sound of the call, turn the screw on the top of the paddle. Also, lightly sand off the chalk so the paddle is clean. Then re-chalk it and you're ready to go.

Quality box calls are investments and will pay you dividends for many years when it's given some regular tender, loving care.

MATT MORRETT
Harrisburg, PA

Matt Morrett is a Grand National Champion caller. As a pro-staffer for H.S. Strut he's appeared in numerous videos and TV shows. In 1998 he called in a 25-pound tom for his dad in Dauphin County. He also hunts in Juniata, Perry and Susquehanna counties.

SHOCK AND AWE: LOCATOR CALLS DO THE TRICK

Like turkey calls, locator calls do not always make a turkey respond. But when they do they're invaluable.

My three favorites are the barred owl, crow and the friction turkey call. Early morning and late evening are the times to get out your barred owl call – unless you can do your own effective mouth imitation, which many hunters can. Roosted gobblers seem to respond very well to the barred owl. Many people think that they are enemies of turkeys but that is not fact. During the mating season gobblers are high strung and ready to show off. Loud noises often draw a response, called the "shock gobble."

With that term in mind, it is important when using any type of locator call to make it a sudden, loud, and sharp sound to startle a gobbler into responding. The most natural times to hear barred owls in nature are in the early morning and at roost time.

Once the crows wake up, imitating their calls becomes a must. However, as a locator call you want to "shock" that gobbler, not sound like you've joined a medley of crows. My favorite way to utilize a crow call is when a turkey has revealed his position I'll check him periodically as I move in and approach a set-up. Gobblers don't always respond to a crow, but when they do it makes our job much easier.

The third locater-type call is cutting on a friction turkey call. The reason I choose a friction call is that it has much more volume than a diaphragm (for most people) and is very realistic.

With the turkey populations growing like they are across the country and especially here in Pennsylvania, we are finding that many days gobblers are quite vocal at daylight, then they fly down with their hens and keep quiet for a time. That means that the hunting gets better later in the morning.

From 8 a.m. until about 10 a.m. is a great time to utilize the method dubbed "cuttin' and runnin." Always start your first series soft, then add volume.

SHIRLEY GRENOBLE
Altoona, PA

Shirley Grenoble is arguably the best known female turkey hunter in the state and a respected outdoor writer. She is a founding member of P-NWTF and served on the NWTF board. She hunts Blair, Wayne and Armstrong counties.

THE CLUCK: EFFECTIVE YET UNDERRATED

The cluck is one of the two calls most often used by turkeys and most easily reproduced by hunters. Yet, it's the one most seldom used.

Nowadays, when hunters fill the woods with fancy cackles, cutts, fighting purrs and the like, the cluck is often overlooked as the persuasive and convincing call it really is. From the day poults are hatched, hens cluck to them. It's the reassurance call that tells the poults "Everything is O.K. here."

Hens and their young cluck constantly to keep track of one another or to ask in turkey language: "Where are you?" When one comes to your calls clucking, that's what he's asking.

Clucks work best to entice a turkey to keep on coming. It often convinces a wary, call-hardened turkey that you are the real thing. He'll discern your clucks from the cacophony of cackles and loud yelps resonating through the woods, especially in the spring. For me, that truth was reinforced in the spring of 2004. After two weeks of chasing gobblers, using nearly every call I owned and every day trying to get a gobbler to not only answer me but to come to me, I decided it was time for a change of tactics.

On my last day to hunt that spring I selected a set-up spot not 500 yards off a well-traveled road and in a hard-hunted public gamelands, determined to hunt the way I knew was most apt to work. I set up in a comfortable position and pulled out my Primos Super-Slate. I mouthed my four-reed diaphragm and began to call, mostly calm clucks. Within 10 minutes a gobbler answered from far off and he came in to my calls. I'm convinced that my everyday clucks and a few yelps convinced him I was really the hen was still looking for. For the record, it was a 19-pound, double-bearded Pennsylvania trophy.

The call can be made on just about any calling device you can operate but it's my opinion that slate and glass calls make the most authentic-sounding clucks.

TOM NEUMANN
Export, PA

Tom Neumann and brother Greg co-own Penn's Woods Game Calls and co-host the Hunting With Penn's Woods TV show. Tom uses his company-made Waterproof Wizard Box Call when patrolling the Westmoreland County turkey woods.

TRY A BARREL-MOUNTED PUSH-BUTTON CALL

Although the push button box call is an excellent choice for beginners, it is also a reliable call for experienced hunters. A gun barrel mountable push-button gives a hunter the advantage of calling with the ability to keep the gun shouldered at all times; important when a turkey is within that critical 50 yards. A simple one-inch movement of the finger will deliver excellent and reassuring turkey talk to an approaching turkey.

The most crucial part of any hunt is when the turkey is less than 60 yards or is visible to the hunter. At this point, there must be absolutely no movement and, if you don't use mouth calls, a gun mountable push button gives a hunter the ability to call while being ready to shoot at the same time. This call is best known for delivering subtle clucks and purrs that in turkey language reassures an approaching turkey that everything is "O.K."

A couple drawbacks to this type of call are the constant maintenance and fair weather functionality. When the call begins to slide and not sound right, try chalking as often as necessary to achieve the desired sounds. Sand the underside of the striking surface but avoid sanding the striker. The bottom of most push button calls is constructed of very thin wood which when wet will ruin its good sound. The Penn's Woods Quick Gun Mount Call comes with a waterproof cover which makes the call 100 percent waterproof, yet delivers excellent imitations.

One other important consideration is to never place the call under or near the holes on a ported barrel. These barrels are designed to reduce recoil by allowing some of the energy from the shot to escape from the barrel and not just the muzzle. Placing the call over, under or near the ports may shatter your call into numerous pieces.

Of course, the same call can be used without mounting it on the gun. It, too, takes minimal movement and is easily carried in a vest pocket.

Whether you're a beginner or expert turkey hunter, don't let the simplicity of any push button call deter you from using it.

PAUL BUTSKI
Scio, NY
Paul Butski is a six-time U.S. Open and three-time Grand National Turkey Calling champion. He owns and operates Butski's Game Calls. When not guiding turkey hunters in Alabama, he hunts New York and Potter and Sullivan counties in Pennsylvania.

TIPS ON MOUTH CALL CARE

The mouth diaphragm turkey call, in my opinion, is the most versatile of any turkey call. However, it is also the most difficult to master. It takes a lot of practice to become proficient.

The call is made of a metal or plastic frame. Latex is set inside the frame with single or multiple layers which are stretched and crimped inside the frame. It is then covered with tape trimmed to form a comfortable fit in the mouth. There are a variety of cuts or notches that can be put in the latex to produce various tones.

Call care is very simple; it consists of a few do's and don'ts. First, do not leave your call on the car's dashboard. Sunlight and heat will deteriorate the latex. Do not run them through the washer. Prolonged moisture will break down the glue on the tape, causing it to lift off the frame. Do not leave your wet calls closed up in containers as this will cause your call to sound flat.

These calls can be used from one year to the next depending on how often you hunt or practice. The more you use the diaphragm call, the shorter life span it will have. Latex loses its elasticity with a lot of use. Do rinse your calls after use. Let them air dry and then refrigerate. This will increase the longevity. When your reeds stick do not pull them apart. This will stretch or even tear the latex and ruin a great call.

Instead, place the call in your mouth and within minutes the moisture will separate the reeds. Be aware that calls just off the shelf can be bad because of shelf placement (light) or shelf life (too old).

It's important to always have back ups and carry a variety of turkey calls. You never know what tone or what call will catch an old gobbler's ear.

ED KEMP
Pen Argyl, PA

Ed Kemp has entered but four call-making contests and took first place in three. He served on the P-NWTF board and belongs to the Pocono Chapter. Ed hunts turkeys in Northampton, Monroe and Pike counties.

CUSTOM CALLS
DRAWING BIG INTEREST

Back in the early 70's when I started turkey hunting the Lynch box call seemed to be the only one on the shelf and not all sporting goods stores carried it. Nowadays there is a call for every situation. But to my knowledge the box call has been responsible for more turkeys brought home for dinner than any other device out there: Which brings up the subject of custom made calls.

Most box calls today can be traced back to the Gibson design patented in 1897 by Henry C. Gibson. Its sound chamber has been copied more than any other. Tom Turpin started making box calls around 1926 and his early calls were inspired by Gibson's creations. A little more than a decade later Mike Lynch started a call business that continues to manufacture them today.

Now there are hundreds of thousands of turkey hunters nationwide and I suspect a good portion of those hunters have thought about, or have made, calls themselves. It's very gratifying to make a call and lure in a turkey with it.

Native Americans were calling turkeys long before anything like today's calls were ever thought about. In the last 20 years or so custom calls have gained big interest with hunters and collectors alike. I believe Pennsylvanian Earl Mickel's books and the NWTF Grand National Custom Call Contest have a lot to do with it.

But there is a difference between a call made by someone to call in turkeys and a call made to win a contest. Most custom calls are made with intentions of being able to call in a turkey, but once they leave the hands of the makers many sit on display. For most of us it would be hard to use a call in the field for which you paid hundreds of dollars.

Turkey hunting roots run deep in Pennsylvania and so does the art of making calls. There have been dozens of call makers over the years in Pennsylvania and a few make a living at it. As for myself, I can't quit my day job yet. Most of my calls are made of ebony and African ivory inlay and given away as gifts.

When asked: What is your favorite call? I say honestly and proudly that it's the one with which my son killed a turkey. That's what this hobby is all about.

ROBBY ROHM
Blain, PA

Robby Rohm owns Rohm Brothers Turkey Calls in Perry County. He holds numerous calling titles and was the U.S. Grand National Champion in 1980. One of his favorite turkey guns is a Knight TK2000 muzzleloader.

ADVICE FOR USING FRICTION CALLS IN THE RAIN

Moisture and friction calls just don't mix.

In any such sort of contact, moisture – rain, snow, sleet or dropping your call in a stream – always "wins." The friction call and the hunter lose.

Humidity and rain can be a real pain when trying to lure a crafty longbeard into range. Most friction calls consist of a few parts that make them run. My choice of friction calls in fall for longbeards is slate. But in damp weather it will absorb water.

When I'm hunting on drizzly days I wear a poncho style raincoat. With this style of raingear I not only keep myself, my vest and other gear dry, but I can also conceal my movement when running a friction style call. If you run a wooden peg on your slate call, try dipping the tip in a two-part epoxy glue, then letting it harden. Follow by sanding it down to where it barely coats the end of the peg. This will keep moisture from penetrating the end of the peg and prevents it from skipping across the slate.

Box calls have parts made of various woods. Any wood not treated properly will absorb water. If I want to use a box call under rainy conditions I wear my poncho and underneath it my vest. When I want to call, I slip my hands inside and pick my dry call. I'll then lean forward a bit and run my box call without obstruction. Some hunters like to run them in a plastic bag, but I find this to be cumbersome and noisy. On light misty days I just lean forward to protect the call with my body. I always carry a zip lock freezer bag with me to store my calls if caught in an unexpected rain or storm.

If you happen to get your friction call damp, dry it slowly when you get home. Do not dry it fast unless you want to invest in another call. When dry, sand the chalk off of the lid with a light sandpaper. Wait a day and re-chalk with only a light coating. Most hunters put way to much chalk on their box calls. Use oil free type of chalk, such as carpenters chalk.

CHAPTER 19

IT'S THE LAW

Know Turkey Hunting's Rules and Regulations

Pennsylvania's turkey hunting rules, regulations and laws may not make for the most interesting of reading matter but no hunter should travel afield without an understanding of them.

SPRING AND FALL SEASONS

- In both spring and fall turkey seasons it is unlawful to possess or use live turkeys as decoys or to use dogs, drives or electronic callers or devices.
- Blinds: The use of turkey blinds is now legal under the following definition: Any artificial or manufactured turkey blind consisting of all manmade materials of sufficient density to block the detection of movement within the blind from an observer located outside the blind.

Artificial or manufactured turkey blinds consisting of all manmade materials means blinds must be constructed of plastic, nylon, canvas, cotton cloth, plywood or other manmade materials. Blinds made by piling rocks, logs, branches and other debris are unlawful. The blind must completely enclose the hunter on all four sides and from above to block the detection of movement within the blind. When fluorescent orange is required at a stationary calling location in fall seasons, at least 100 square inches must be displayed outside the blind and within 15 feet of the blind. It must be visible in 360 degrees.

Legal Arms (Single projectile ammo is illegal in spring).

- Manually operated rifles and handguns.
- Manually operated and autoloading shotguns limited to a three-shell capacity in the chamber and magazine combined.
- Muzzleloading shotguns, rifles and handguns.
- Long, recurve or compound bows with broadheads of cutting-edge design.

Ammo and Arrows

- Single-projectile ammunition or shot no larger than No. 4 lead, bismuth-tin, or tungsten-iron and No. 2 steel.
- The fall season in Wildlife Management Units 1A, 1B, 2A, 2B, 5C and 5D is limited to shotguns and long, recurve or compound bows with broadheads of cutting-edge design. Crossbows are not legal for turkey hunting with the exception of authorized crossbow permit holders.

SPRING GOBBLER SEASON ONLY

- Rifle-shotgun combinations may be used if ammunition is limited to shotgun shells. Carrying or using single projectile ammunition, rifles or handguns is unlawful.
- Only turkeys with visible beards are legal.
- Hunting is by calling only – no stalking – from one-half hour before sunrise until noon. Hunters shall be out of the woods by 1 p.m.
- While moving, a minimum of 100 square inches of solid fluorescent orange material visible 360 degrees (an orange hat suffices) must be worn on the head.

FALL SEASON ONLY

- Turkey hunters must wear at least 250 square inches of fluorescent orange on the head, chest and back, visible 360 degrees, when moving. When at a stationary calling location, the orange may be removed providing the hunter places at least 100 square inches of fluorescent orange, visible in all directions, within 15 feet of the calling location, and visible 360 degrees.
- An exception occurs in Wildlife Management Units limited to hunting with shotguns and bows and arrows (WMUs 1A, 1B 2A, 2B, 5B, 5C, and 5D). Hunters, when moving, must wear at least 100 square inches of solid fluorescent orange on the head. Even though fluorescent orange is not required in these zones at stationary calling locations, it is strongly recommended.

TAGGING AND REPORTING

- Successful turkey hunters must follow all instructions printed on tags supplied with licenses. Separate tags are provided for fall and spring seasons. The turkey must be tagged immediately after harvest and before the carcass is moved. The tag must be securely attached to a leg until the bird is prepared for consumption or mounting. Once you have used your tag it is unlawful to possess it in the field.

Glenn Lindaman of Whitehall fills out a tag for his gobbler.

- Turkey hunters must report harvests to the Game Commission in Harrisburg within 10 days, using a postage-paid report card supplied with each hunting license. In addition to other information, hunters are asked to identify the WMU in which the bird was taken.

Author's note: Laws, rules, regulations, seasons and changes in WMU requirements may vary from year to year. The regs listed here were valid in the 2004-2005 seasons, as this book was being written. They should serve only as general guidelines. For updates, check the annual "Pennsylvania Hunting & Trapping Digest" issued with each year's license purchase.

On the Hunt: The Eternal Vernal Season

Patience and perseverance sometimes produce rewards such as this.

CHAPTER 20

SETTING THE SPRING SEASON

Too Early, Too Late or Just Right?

The 35 acres of deer and turkey woods surrounding my Northampton County home bring daily blessings, not the least of which are the late winter and early spring sightings of hens moving through or gobblers testing their lungs as a tune-up for the upcoming spring while standing ankle-deep in snow.

For some hunters, it's hard to resist the urge to test a few calls on the chatterbox birds. Other hunters curse such behavior as, they insist, it will make the toms more wary by the time the season opener arrives.

"Understandably, turkey hunters want to be out in the woods when they believe that the birds are gobbling best,"

Gobbling activity can be heard long before the late April opener.

says Bob Eriksen, Pennsylvania biologist for the NWTF. "But when is the most gobbling going to take place each year and how should our hunting season be timed?"

"In general, the season should open when hen turkeys are apt to be laying or incubating their eggs," said Eriksen. "In Pennsylvania there is a 6-week window of opportunity for the spring season during April and May."

When Gobbling Peaks

Breeding and nest initiation are stimulated by the length of daylight hours, called a photoperiod, but weather also plays a role. When light conditions are right for breeding, cold or unseasonably warm weather can speed up or delay the process by as much as two weeks.

Biologists use several methods to determine when the spring season should occur. According to the research results, there are three peaks of gobbling activity:

The average Pennsylvania incubation period begins April 28.

the first associated with the break-up of winter flocks; the second with the beginning of incubation among hens; and the third with re-nesting of hens that lost their first or second nest attempts. Seven to 10 days later many of the hens are incubating as gobbling frequency increases with   toms seeking the few remaining available hens.

Surveys of gobbling toms in New Jersey, Virginia and West Virginia indicate that the second peak occurs from the last few days of April through mid-May. The timing of peak gobbling may vary, based on weather and other factors, but it seldom shifts more than 10 days in either direction.

Spring Season "On Time"

Radio-tagged hens can be closely monitored, allowing researchers to accurately obtain nesting data. According to a 1953-1963 Pennsylvania study, the average statewide incubation date for hen turkeys was April 28. More recently a Game Commission radio-telemetry study on South Mountain in southcentral Pennsylvania revealed that the average incubation date for adult hens was May 8. Juvenile hens in the same study had an average incubation date of May 13.

Both the older data and the recent information suggest that Pennsylvania's spring gobbler season is timed correctly, based on the birds' reproductive cycle. Setting the season carefully is the primary reason we can hunt gobblers during the spring and not affect long-term population levels or disrupt the breeding behavior.

That generally means hunting seasons open around the same time hens begin nesting in most states. In Pennsylvania, hunting occurs when most of the breeding has been completed and many hens are on the nest.

Opener Early as Can Be

Years back, in response to hunter requests for an earlier spring season, Bill Drake, a now-retired PGC biologist, looked at data available from surrounding states. His report recommended that the spring season here open on or near May 1.

Some hunters, however, would like to see an earlier opener. Quiet gobblers, "henned-up" toms, warm days and the thick-leafed woods the final couple weeks are often cited as reasons the spring gobbler season should start earlier.

Interestingly enough, interference from hens may suggest that the season is as early as it can be. Gobbling activity decreases with increases in hunting pressure so the woods may be silent; not because the birds are "gobbled out" but because they have been pressured. An earlier opening day might allow hunters to take advantage of cooler weather, fewer insects, fewer leaves and a time period in which some gobblers might be more susceptible to being called.

But it's hard fact that gobbling activity will always decrease once hunters enter the woods.

Biologists must also consider more than dates and hunter success when setting spring seasons. A later opening date provides a measure of protection for turkey nests, hens and even jakes. Hens that are in the process of laying eggs are prone to abandon their nests if disturbed. Starting the season close to May 1 may reduce nest abandonment. Hens that are not yet incubating are often found with gobblers early in the season. Lone hens in the process of laying eggs are apt to be moving around the woods during hunting hours if the season occurs too early.

The later the season opener, the fewer hens will be available for accidental or illegal shooting. Later opening days may, therefore, enhance the survival of hens.

Illegal Hen Kills

Another consideration is the illegal harvest of hens which, surprisingly, accounted for

Gobbling activity always decreases when hunters enter the woods.

34 percent of all the hen mortality in the spring in Virginia and 13 percent in West Virginia. In 1999-2001 Pennsylvania study showed only a five percent hen mortality from illegal shooting. Jakes tend to be more vulnerable to harvest earlier in the season. As May progresses, hormone levels and testes sizes decrease in jakes more quickly than in adults. Therefore a later season may protect some jakes and allow them to mature. It may also make hunting a bit more difficult because full leaf-out limits visibility and hearing. Pennsylvania has a large turkey population but also a lot of turkey hunters. Tradition and demand for fall hunting opportunity remain high. Given the large number of hunters and the desire to provide good hunting opportunities in both spring and fall, Pennsylvania should benefit from its "late season" debut.

Author's Note: *Information for this chapter was taken from a report on setting the spring season by NWTF biologist Bob Eriksen.*

CHAPTER 21

YESTERDAY'S ERRORS YIELD TODAY'S HITS

Recognizing Mistakes Can Make You a Better Hunter

The barred owl can be a gobbler hunter's best friend when it broadcasts its familiar call.

Goofing up, a friend once proffered over a pre-hunt breakfast, simply makes you smarter.

"If that were the case," I muttered, slurping a maple syrup-soaked hotcake, "I'd be the most intelligent turkey hunter in the country."

Yes, we all make mistakes. But sometimes one day's error becomes the next day's home run. I thought of that on a recent April following a farm country hunt with Matt Morrett, a fellow Pennsylvanian and a full-time pro staffer for Hunter's Specialties. In the dark we'd hiked a quarter-mile from the truck, then hoofed it along the edge of one of many woodlots bordering a complex of soybean and corn fields, arriving at our listening spot at first light.

Morrett needed only a lone barred owl call as gobbles echoed from here and there across the flat countryside. The closest symphony came from a patch of woods not 300 yards to the west. It was the sort of morning when you knew enough opportunities would present themselves to account for full tags well before the noon closing time.

We chose a set-up site along a grassy peninsula jutting from the wood's edge into a field, surmising that the gobblers would fly down, then move out into the open field to feed and strut for the hens we heard yelping in the treetops. Morrett, the designated caller, pressed his back against a tree some 15 yards behind me, offering me first shot and hoping for a good view of the action come playtime.

It wasn't long before the fluttering of wings was heard as birds pitched down. The first to appear in the yet-to-be-plowed field were three hens. Behind them were as many gobblers, two of which were strutting and offering an occasional gobble at Morrett's clucks and yelps. But with jennies and adult hens in abundance, the toms expressed no desire to find the "invisible" hen.

As luck had it, however, only one of the 15 birds which eventually appeared was close enough for a shot, and that loner was a hen. Each, as if it had been rehearsed, played follow the leader on the far side of the field, moving off at an angle as they began their day's travels.

For three hours we watched as the troop fed together as far as 300 yards across the barren field, picking up and losing straggler hens and jakes from time to time. We stayed in position, watching through binoculars as the birds fed and bred.

Then, finally, as if someone had given an order, the entire flock pulled anchor and headed back in our direction. Within minutes a pair of hens walked by, a few yards from where I'd watched the closest bird pass at sunrise. My interest, however, was focused on four longbeards following her. My Remington 1100SP choked tightly with a Hunter's Specialties "Undertaker" choke tube and loaded with 3-inch Winchester Supreme No. 5 shot was braced atop my knee. All was ready for a tug of the trigger. Matt had offered several soft yelps, then became silent, satisfied to let nature take its course.

The birds were marching in and there seemed no reason to do anything but wait. From the corner of my eye I noted a longbeard standing alone at about 30 yards, its neck stretched high, seeking Morrett's subdued calls. Instead of swinging on him, however, I chose to remain frozen and concentrate on the toms moving into my sight line.

About then, when all seemed to be coming together, trouble showed up in the form of a curious jake who stood directly in front of me at 20 yards while the older gobblers gathered behind him, curious as to what had grabbed the short-beard's attention. Having heard Morrett's calling, the stubborn shortbeard felt he needed to see where "she" was hiding and he wasn't about to move until he did just that.

While Matt awaited my shot, I tried to figure out how to shoot around the jake. Had I squeezed the trigger, the shot pattern at

It doesn't happen every time out but every so often gobblers come in together, making it impossible not to kill or injure all of them with a lone shot.

35 yards might have missed the jake but not the longbeards behind him. Clumped in a black mass and a tangle of red necks, the quartet seemed intrigued by the jake's rapt attention. Had I shot, I'd have probably have killed every one of the birds, a sure state record – and hefty fine.

To complete my tale of woe, the birds ducked their heads and disappeared behind several locust trees, then melted into the same woodlot they'd spent the night. Soon all was quiet with nary a black feather to be seen. The morning hadn't been the sure thing we'd anticipated.

The next day, though, took a different turn. Betting that the birds would roost in the same trees and travel the same routes into the field, Morrett sat against the tree I'd used the previous morning. I traded mine for another small tree (a spiny locust, unfortunately) some 40 yards closer to the wood's edge and on the very end of the field. Less than five minutes after flydown the hunt came to an anti-climatic end as a 21-pound tom with a 10-1/2 inch beard paused to stretch its wings 25 yards away. It all happened within 20 minutes following Morrett's first wake-up calls.

Our previous day's failure had resulted in this day's success. We'd resolved the errors of our ways. The lesson of using today's mistakes to tomorrow's advantage was driven home for the "umpteenth" time in my 30 or so years of spring gobbler experience.

That afternoon Morrett and I got into a discussion about the many now-humorous foul-ups we've made over the years that eventually turned despair into good fortune. Turkey hunting lessons, as with life itself, come with a price tag but have been priceless over the years.

Just because he's looking your way doesn't mean he sees you.

Here are a few of them.

- **The eyes have it**

Error number one in my turkey hunting career came only a few hours into a South Carolina hunt in 1978. The pair of gobblers my longtime friend and fellow Pennsylvanian Bob Clark called in during a marathon session finally came into view, then came over a rise and paused at 35 yards to stretch their necks and view the new scenery. Being inexperienced, I thought the lead tom had stretched its neck because it saw something it didn't like ..... namely me. I hastily fired, missing on the first shot but, thankfully, hitting the mark as the bird flew off. Since then I've been perused by dozens of birds, many of which came into range "on alert" and went home over my shoulder. Now, however, I know enough to have faith in my camouflage and the patience to sit tight.

Patience and perseverance are qualities every turkey hunter needs to develop.

• Back in the saddle

It took several years of screw-ups but I finally learned that saddles (landforms, not horse seats) are excellent places to set up. A saddle is simply a low area on a rolling hillside paralleled or circled by higher spots. It's in these environs that I like to set up shop, even though the view may be limited. The hunter, however, is usually hidden by such depressions and the first sight one may get of an approaching gobbler is its white head on the horizon. Saddles are also good sites for "blind calling" to vocal birds. The lesson was driven home more than once by the turkeys I'd bumped from saddles over the years. If you can get a bird to gobble, the advantage goes to you as you know he's on the way.

• Heads up

Thoughts of the first turkey we ever missed and subsequent gobblers that went their merry ways after close calls all come back to haunt us now and then. Maybe it's because we knew exactly why we missed only seconds after the quarry made good its getaway. Most misses, I've learned, come from shooting over a turkey's head. Why? After reviewing the episodes in my mind and on the shooting range, I realized that the front bead was in plain sight and aligned with the gobbler's neck (as detailed in a previous chapter) but it wasn't in line with the rear sight. Hence, I shot above the incoming turkey.

• Only five minutes more

Everyone recalls the first time he or she worked a bird for an hour, until it fell silent, then pulled down the mask, removed the gloves and stood only to hear the maddening "putt" and see a black form sprint through the woods and out of sight. Perhaps it's because I've grown older and tire more easily that I have also become more patient. I can recall only one case of chasing off a gobbler that worked to a call, and then developed lockjaw, in the past dozen or so years.

Of course, you must move sooner or later but I no longer do it before giving the bird every chance to make a mistake. When I'm not sure if he's departed or is simply biding time behind a saddle or deadfall nearby, I'll talk myself into waiting another five minutes ... or more.

• **Power gobblers**.

Many places I hunt both near and far from home – public and private lands – have powerlines running through the properties. For many years I ignored them, except at deer hunting time. But why not use them at gobbler time, too? A decade or so ago I developed an affinity for powerlines after sighting a gobbler from some 300 yards using a binocular. I then closed the distance in the confines of the adjacent woods, and eventually called the lone tom close enough for a shot.

At times when there's a need to cross open areas, wait to call until you arrive at the wood's edge.

Find a high spot and glass the open area. If a bird is seen, pick out landforms to mark your way, then make your move inside the wood line, cautiously poking your head out now and again to get your bearings. Don't call until you know you're within 100 yards or so of the bird, unless you can no longer see him.

• **Bed and breakfast**

One of the more convincing lessons I learned in the turkey woods came courtesy of late and legendary Ben Rogers Lee.

"In the middle of the day turkeys will go back to where they spent the night," Big Ben advised one hot, April afternoon in a southern Alabama woods. "You get back there and sit tight and they'll come by." So, we "got back there," sat, and within 20 minutes Ben brought four jakes to within shotgun range. We passed on the young toms but the grand master of turkey hunting was satisfied that he'd made his point. A day later Ben claimed to have killed a longbeard in similar fashion.

• **Rainy days are field days**

I received my turkey education from several friends, a couple who harbored the belief that if it's raining you can forget about getting a gobbler's interest.

"We might as well stay home until the sun comes out," one of them told me one drizzly, foggy morning as we sat in his kitchen drinking coffee.

So we did. In my travels here and there across turkey country, however, I often saw birds in fields on rainy days. Yes, some of them looked more like forlorn turkey vultures than wild turkeys as they moped, dripping wet, beneath wide-crowned, field-edge trees. I later learned that turkeys were not only visible but also "callable" on rainy days. I'd made the mistake of believing my hunting buddies who never went out themselves when it rained so how would they know what turkeys did when the skies opened? Today, my Gore-Tex and I don't let a little shower deter our turkey time.

- **All choked up**

Yet another error of my ways came in Alabama a few years back when Bo Pittman of White Oak Plantation called in four longbeards to 18 steps and I missed a shot at the lead bird. Instead of the pattern hitting the neck and head, however, the tight ball of lead pellets cleanly severed a sapling at nine yards, as videographer and guide Tes Jolly discovered when we tried to figure out what had happened. The choke, which would have done the job at 30-40 yards, simply held too dense a pattern at nine yards yard and left my target bird unscathed.

The lesson here is that I never gave sufficient thought to the tight grouping a 3-inch magnum and an ultra-tight choke provide when the target is close and you need to fire into thick quarters, such as the saplings through which the birds were traveling. The next time you pattern your gobbler gun, also take aim at a 10 yard target and note the devastating yet small pattern a choke tube can create. That knowledge might one day come in handy while afield.

- **Wait until later**

Pennsylvania is what I call an "opening day" state. Hundreds of thousands of sportsmen and women celebrate trout, turkey, small game and deer traditions by heading out on season debuts. It took me many years to comprehend that the turkey season opener and the first weeks of the season aren't necessarily the best. Sure, I'll be out there with everyone else, but it's the final weeks of the season when the hens are nesting that I most enjoy if my tag remains attached to my license.

"If I had to give up all but three days to hunt turkeys, I'd choose the last three days of the season when everything's thick and green," Morrett tells the audiences to whom he speaks.

"By then the gobblers have lost most of the hens and may be roosting alone. In fact, they'll stay in the roost longer waiting for hens. I've bumped them off roost trees long after first light during that last week of the season. Besides, you'll have most of the woods to yourself."

- **Don't think too much**

No matter if I'm after bucks or gobblers, I know that I'm sometimes my own worst enemy because I think too much – often based on long-past happenings

*To some spring turkey hunters the last two weeks
of the greening season are the best.*

which hadn't turned out with the weight of a bird on my shoulder when it should have. Morrett admits to catering to the same gremlin.

"I know sometimes I try too hard to make things happen that just won't happen," said Morrett. "I'm especially guilty of doing it when I guide someone and badly want him to get a bird."

Getting into such pressure situations often triggers responses that do not reflect wise decision making.

"I've learned many times that trying too hard on one particular bird will bring nothing but frustration and the next day you go out there and it's a piece of cake," Morrett reflects. "It's nothing that you did wrong. It's just the way turkeys can be. My advice is to sit back, relax and give it your best shot. It's not brain surgery. It's turkey hunting."

It all boils down to this: Don't become discouraged when some foolish error or the wrong decision sends you home "birdless." Instead, analyze what went wrong and, calling on past experiences, turn yesterday's errors into tomorrow's hits.

RAY SMITH
Jersey Shore
Ray Smith is the owner of River Valley
Game Calls and River Valley Outfitters.
In addition to Sullivan and Lycoming
counties, he also guides in Texas and
Kansas. Ray serves on the board of
directors of P-NWTF.

NO RULES:
JUST EDUCATED GUESSES

Turkey hunting success can be summed up as "a mix of educated guesses." This is what challenges us to pursue this creature that has the ability to make us crazy. Often times I find myself sitting high on a ridge when the turkeys are not vocal, leaving me with several choices; sit it out and hope turkeys slip into my location or cover some ground hitting the creek bottoms or valleys.

Most of my decisions are based on the situation, habitat and the time of the season that I am hunting. Most mornings find me located above where the turkeys will be found. By experience I've come to the conclusion that turkeys are much easier to call uphill or on the level, not downhill. Turkeys are more leery to come down to an area that they can see so well, as if they know they are making a mistake.

When starting your hunt atop a ridge your calls will carry much farther then if you were in a creek bottom. This also gives you the opportunity to hear turkeys at a greater distance, enabling you to pinpoint their location and make a move if you have to.

That said, keep in mind that turkeys do not have a rulebook. Lessons are learned by trial and error. There is not a rule that says staying in one location will bag you a bird. Nor does it say that moving all the time will help. Let the turkey dictate to you what they want.

In fall, turkeys are always on the move. The successful hunter moves with them. In spring, turkeys tend to stay in the same area day to day. You should stay there for obvious reasons.

The wonderful thing about turkey hunting is you can hunt them successfully by being either passive or aggressive. It's your choice.

As said, "There are no rules, just a mix of educated guesses"

JIM PANARO
Ebensburg, PA

Jim Panaro is president of the
Chesquehanna Chapter of P-NWTF
and has 25 years of turkey hunting
experience under his boots. He frequents
the woodlands of Cambria, Indiana and
Huntingdon counties.

ANALYZE YOUR SET-UP,
THEN BLEND IN

People often think that the most successful turkey hunters are the best turkey
callers. But in my experience the hands down best turkey hunters are those who
have mastered the set-up and know how to blend in with their environment.

The classic method is to sit in front of a tree wider than your shoulders to
break up the human outline. This is an important safety consideration as well as
protection from hunters who may be trying to sneak in on the "turkey" they hear
calling and inexcusably taking a shot at movement.

But what do you do when you encounter a turkey in agricultural areas, heavily
timbered woodlands, or one of the reclaimed surface mines that are common in
my neck of the woods? There, there's not a big tree to be found?

Adaptability is the key to these situations. You have little choice but what's
available as a backstop when a big tree isn't handy. Round hay bales in farming
areas or drainage ditches and brush piles around the edges of reclaimed mine sites
work well. Whatever you choose it is absolutely critical to remain as motionless as
possible and have confidence in your complete head-to-toe camouflage. That only
works when movement is kept to a minimum, of course.

The biggest "secret" I use – no matter where I'm set up – is to worry less about
what is behind me and more about what is in front of me. I try to set up with some
sort of small diversion about 35 yards between the turkey and me. This diversion
can be the brow of a ridge line or mountain bench, a curve in a logging or gas well
road or a downed tree or thicket. The variety is almost endless.

As the situation permits, I like to walk up to the "obstacle," then turn around so
that I am looking from the turkey's point of view. Then I'll finalize my choice of a
set-up site. The idea is to keep the old gobbler guessing at just where the "hen" is
calling from and let his curiosity get the best of him.

If the bird hasn't spotted you and your gun's in position, don't be too hasty.
But if he starts to get nervous or jittery the show's over and you'd better be
ready to tug the trigger.

ERIC BAKER
Port Matilda, PA

Eric Baker began hunting at age 12 and is a current member of the Primos Game Calls pro-staff. He hunts primarily on public lands and visits three to five other states each spring.

ALL THE RIGHT MOVES

A longbeard suddenly appears within effective shooting range but your gun is not aimed in its line of fire.

Now what?

If you've hunted turkeys, I know you've found yourself in this situation. My advice? Simply do not lose your cool. The worst thing you can do is rush your shot. Doing so will likely result in a miss or, even worse, a wounded bird.

Instead, assess the situation to determine if the bird will likely walk behind anything within your effective shooting area (the area covered by swinging your gun left and right), such as a large tree or deadfall. If such a blind spot exists, wait for the best opportunity and make only deliberate and measured moves. If the bird is strutting, wait until he faces away and use his fanned tail to make your move.

What if there is nothing to preoccupy or otherwise "hide" a bird responding to your calling?

In my opinion you only have two options. The first is to make your move at a snail's pace. The farther away the bird, the better the opportunity to attempt slow, brief movements which will not be as noticeable at 30 yards as at 15 yards. Assuming you're totally camouflaged and move slowly and timely, you'll be amazed how much you can move, particularly if the bird is walking.

The second option is to make a fast, deliberate move to get on the bird and shoot before he starts to run or fly. Although he will not wait long, when alarmed a turkey will often hesitate for a split second before making his departure. This hesitation can be just enough to allow a quick shot.

Be forewarned that option two comes with more risk than option one. In my opinion the shot should not be taken if the bird starts to run or fly due to the great possibility of wounding him. But it can be done at close range.

Of course, there will be times when nothing works and the bird makes a clean getaway. But then, that's part of turkey hunting.

Spend a bit of time before your next hunt practicing how you should move when the situation presents itself.

SCOTT BASEHORE
Denver, PA

Scott Basehore is a custom box call maker who prefers his cedar paddlebox to any other. He's been turkey hunting for 30 years and spends most of his turkey time in Potter, Lebanon and Sullivan counties.

HUNT THE BREAKS FOR SILENT TOMS

The most important part of being successful in turkey hunting is scouting – knowing where the birds roost and, more importantly, where they go throughout the morning. Scouting is especially important on those mornings when the gobblers have developed lockjaw. Then I like to set up in areas that hold a lot of sign, like scratchings, droppings, or visual contact.

With Pennsylvania's expanding turkey flock there is more competition with hens. A lot of contented gobblers have hens with them the entire season. This can put a damper on gobbling activity. By finding areas the turkeys favor, this can greatly increase your chances of putting a tag on an old longbeard.

When setting up, I like to find some sort of break in the terrain so that the turkeys can't see a long distance to where the calling is originating. But don't be so well hidden that turkeys will be right on top of you when they show up. When calling from these set ups, I favor a more subtle approach.

I like to do a lot of soft clucking, purring and yelping; all contentment calls. From time to time I may perform a short series of cuts. I can't count the number of times I've sent out cuts only to have a bird fire back from close by. Often he'll come in, either by himself or following several hens. I have used this tactic on all subspecies of turkeys, not only Easterns.

This type of hunting requires a fair amount of patience and being still. With turkey blinds now legal in Pennsylvania, it makes remaining undetected and moving your gun or bow much easier. When I set up in one of these areas I usually plan to stay two to three hours. Take along a snack and something to drink and be ready. You never know when that old boy may show up.

A final note: This type of hunting produces a lesser chance of bumping non-vocal birds as opposed to walking and calling. The latter may put birds on alert and further compound the problem posed by any lack of gobbling activity.

RON BROTHERS
Sellersville, PA

Ron Brothers serves as president of the Southeast Silver Spurs Chapter of the P-NWTF. His biggest gobbler is a 26-1/2-pound Kansas Eastern with 1-5/8 inch spurs. Back home, Ron hunts Bucks and Bradford counties.

PATIENCE CAN MAKE ALL THE DIFFERENCE

I've been hunting turkeys since the early 60s and if you were to ask me what lesson I've learned that's helped me the most I'd say it in one word: Patience.

Sitting that extra 15 minutes has often yielded a bird I didn't know was there. Many hunters lacking patience tend to move to soon when they might have been better served by simply staying where they were a bit longer.

When I was first learning to turkey hunt the legendary Ben Rogers Lee was the only one out there with a cassette tape on turkey hunting. One of the things he said has stayed with me all these years: "When you call to a turkey he knows so well where (the sound) is coming from that if you were in a hole, he would come and look in that hole."

Of course, turkeys don't crawl into holes but if they could, they would. That has proven to be true many times.

An example is a trip to Mexico where I attempted to bowhunt for the Gould's subspecies from a blind. With me was Joe Hall, Jr. of Hally Callers in southeastern Pennsylvania. He was there to film me. We set up and relaxed and the next thing we both dozed off. Somewhere in the cobwebs of my mind we both heard a cluck. Opening our eyes, we were shocked to see a jake looking in our blind.

Had we decided to leave rather than stay, that wouldn't have happened.

Sometimes when you're working a bird he will shut up and 10 or 20 minutes later you hear a sound in the leaves behind you. The hardest thing in the world to do is not to look to see what it is, even if you believe it is only a squirrel. If it's the same gobbler working in behind you, resist the urge to swing and take a wild shot before the bird takes off running or flying.

I choose not to ever try to quickly turn and take a snap shot. Instead, stay motionless and hope he'll walk to either side of you. If he walks away unaware, try looping him in the direction he walked and attempt to call him again from a different spot.

CHAPTER 22

EVERY DAY'S A FIELD DAY
Fair or Foul, Hit the Wide Open Spaces

Power lines are excellent sites for locating turkeys by using binoculars and studying the near and distant clearings.

Turkey hunters are well aware that drizzles and downpours will drive wet and wild turkeys into fields and pastures. But that's not the only time it may be better to stick to the wide open spaces instead of tromping through the woods.

I learned that lesson from Larry Norton, a well known southern guide who occasionally comes north to hunt Pennsylvania turkeys.

"Ready to walk?" Norton asked as he drew his truck to a halt along a muddy road in mid-morning. It was my first hunt of the year and I was primed, in anticipation of hearing the first gobble of the infant season.

Norton, with numerous calling titles to his credit, casually remarked that hunting open spaces – fallow fields, croplands, power lines and gas pipelines – is one of his favorite ways of hunting; not as a back-up venture but as the focal point for locating gobblers by sight.

"You've got to remember that turkeys are creatures of habit," advised Norton as we walked toward a backwoods power line cut.

"They'll be out in fields at one time or another every day. Checking them out by sight instead of sound reveals the direction they're heading and that's the key to success at places like this."

It sounded easy enough.

As we approached the power line a mile or so off the gravel road, Norton slowed his pace. Kneeling, then lifting a binocular to his eyes, he scanned down the 150-yard wide opening allowing us visibility for more than a mile. I followed suit, soon spotting some black "dots" a half-mile or so off through my 10x binocular. Despite the distance, it was obvious two of the birds were gobblers, taking turns strutting their stuff to the largely disinterested hens, which were slowly – painstakingly slowly – moving in our direction.

Had I been alone, I'd probably have worked a bit closer under the cover of the woods, set up along the wood's edge, issued a few casual yelps or feeding calls, then sat tight just inside the treeline, gambling that the birds wouldn't exit the field before they got to me.

Had I been alone I'd probably have gone back to camp empty-handed.

Norton had a different plan in mind. He knew something I didn't. Between us and them was an attraction even Norton's award-winning calling might not be able to overcome. The birds were near a hardwood bottom roosting area which Norton had discovered on a previous visit. Then he'd watched from afar as turkeys cut into the woods on their unhurried ways back toward their sleeping quarters,

Like the hunters seeking them, gobblers often come to fields to check for activity.

even though it was nowhere near roosting time. But he also knew that turkeys often return to their familiar roost tree regions in late morning.

Carefully crossing the grassy field, we moved just inside the treeline and started our trek toward the birds, occasionally working to the field's edge to try to spot their progress. Fifteen minutes later we were where we wanted to be, having closed the gap by at least 400 yards. Yet, we still were not quite sure where the birds had gone as the rolling right-of-way and some high grasses blocked our views. Norton called only occasionally and softly, choosing to fool the longbeards into believing that a few new hens were on the way.

The author took his 2001 gobbler from a wheat field in Northampton County.

Finally, we set up side-by-side about 30 yards inside the treeline and Norton began his concert. But to no avail. His turkey talk seemed to fall on deaf ears. Finally a crow screamed and one of the birds sounded off, probably from less than 150 yards off. Calls carry farther in the wide open spaces and it's often hard to pass judgment as to just how far away a talkative gobbler may be.

Larry used the cue to broadcast several loud yelps, followed by soft feeding calls and leaf-scratching, leaving the next move to them. Fifteen minutes later a black form emerged from the edge of the field, stretching its head high as Larry again purred and brushed the leaves at his side. Two minutes later we were kneeling over an 18-pound tom holding a 9-inch beard.

"Every one of these is an education," said Norton, fanning my gobbler's tail.

"Just because a bird doesn't gobble doesn't mean he didn't hear you or isn't interested. You can usually see what he does when you call. You don't need to guess. Seldom can you do that in the woods."

"That's why I love field hunting," he continued.

"One day a bird will gobble his head off every time you call and the next day he won't make a sound. But he can still be called in and you can see how he reacts and move on him. Just knowing for sure that he's there makes it easier to decide what to do. In the woods, you'd probably get up and leave if he didn't answer – and walk away from a bird you could have eventually shot."

Field Days

Yet another star of the turkey hunting world and a specialist in hunting large, open fields is call manufacturer Preston Pitman, who's demonstrated his calling and "strutting" talents on The Tonight Show and the David Letterman Show. Afield he's a conservative caller who likes to watch gobblers in the wide open places and learn from their behavior. He knows just where to set up and has the patience of a rock.

We did just that one sunny morning when, at first light, we stood on the rim of a 200-acre cow pasture and awaited the sound of gobblers' wake-up calls. They were delivered on time and in the half-light of dawn we moved along the edge of the woodline, then set up on a point jutting out into the greening field.

There we would spend the next three hours, another decision I'd probably not have made had I been alone.

As often occurs, the minute the gobblers hit the ground they developed lockjaw, refusing to acknowledge Pittman's clucks and yelps. But as the sun pushed above the treeline, the small flock – a jake, a longbeard and a trio of hens – moved along the edge of yet another point of woodland jutting into the pasture. The gobbler offered a few feeble answers to Pittman's yelps and cuts but gave most of his attention to his hens. Most times they were too far off to hear but through a spotting scope we could see his head thrust forward and his tail fold as he gobbled.

"No problem," Pittman muttered, pressing his back against a nearby oak and closing his eyes. "We'll just wait him out. He'll come around sooner or later to look for what he figures are hens."

And we did.

In the next couple hours we each enjoyed several cat-naps, taking turns in cautiously crawling five feet to a nearby oak, then standing and scanning the field to catch sight of the birds, which at times were mere specks to the naked eye.

By 9:30 I was beginning to get the idea. The flock had now closed to about 200 yards and was working toward the woodlot along which we'd passed three hours earlier. Pittman knew it was time to declare his presence by breaking loose with a series of "fighting purrs," interrupted with an occasional vocal gobble to challenge the "boss."

The ploy worked and for the next 15 minutes the foursome moved closer to our position, picking seeds and insects and regularly gazing our way to try to sight the hidden "hen."

At 35 yards Pittman issued several clucks and the gobbler lifted his head for the last time. The two-year-old, 18-pounder sported sharp 5/8th inch spurs but made up for it with a bushy 10-inch beard.

"If there are two things to learn from this morning it's that you need to set up on the points of woods and you must have patience," said Pittman. "Going to the end of such points permits you to see a whole lot farther than setting up inside the woods."

"You've got to have confidence that when turkeys get into a field they pretty much know where they're going, unless something scares them. And you're just not going to hurry a gobbler if he has hens. If he's not interested in my calls, I shut up and let nature take its course. I won't call until I think I can get his attention.

On rainy, windy days, flocks often leave the woods to feed in fields.

But I'll always stick with birds that I can see. That's the biggest advantage of field hunting. You KNOW the bird's there because you can see him a long way off."

"And that," he concluded, "seldom happens in the deep woods."

Field Hunting in Foul Weather

Fields and power line or gas cuts are also magnets on foul weather days. Turkeys don't like the distractions of wind, rain, leaves blowing and strange sounds they experience in the woods. So they head to a more favorable place, such the edges of fields.

In fact, much of what they do during any bad weather day is determined by the weather. On dark, cloudy mornings they'll linger in trees rather than flying down at normal departure time. I've had roosted birds drop from their perches and, if mist or fog sets in, fly right back up again until the conditions change.

Turkeys will often linger in their roost when weather conditions are poor.

One NWTF study revealed that gobblers will be less likely to sound off if the dozen or so hours prior to flying down were marred with wind and/or rain. Bad weather – including early snowfall in the fall season – will also discourage hens and toms from carrying on their normal feeding patterns.

I – as many other turkey hunters – have always been intrigued by near-daily "patterns" that may stretch across a county or much of the state. How many times have you compared notes with other hunters who all reported that "the birds weren't gobbling this morning" when they were hunting 20 or 30 miles away from your hunting spot?

One explanation may be taken from a waterfowl study which revealed that birds

A trio of hardy turkey hunters head out along a Harrisburg-area railroad bed on a foggy morning in 1977.

possess numerous air sacs which are sensitive to barometric pressure. As weather fronts move in, pressure decreases and the birds inherently do some hasty feeding but little else – including "talking."

My view of turkey activity and pending rain is similar to my deer hunting observations: They'll both put on the feedbag when a low pressure weather front is on the horizon. Get there before the storm or rains hit and you might be surprised at the activity. If thunder accompanies an anticipated front, however, gobblers will find it difficult to resist a response.

I always carry raingear at such times, knowing that a turkey's plumage is as efficient as your rainsuit and a slight drizzle will not deter then from filling their crops. As long as I'm warm and dry and the rain isn't of deluge proportions I'll sit it out in a treeline or woodlot edge bordering a field or power line cut, then hope for a hungry and wet flock to appear out in the open.

Decoys offer certain benefits in any type of field hunting.

GLENN LINDAMAN
Whitehall, PA

Glenn Lindaman owns BuckWing Products and has created a variety of helpful and unusual gear for sportsmen, deer and turkey hunters in particular. His new and realistic Bobb'n Head decoy line will hit the catalogs and shops in plenty of time for the 2005 season. Glenn hunts in Pike and Lehigh counties.

REALISTIC DECOYS:
SILENT MIMICS FOR VISUAL ENTICEMENT

Decoy set-up is extremely important for fields and open woodlands. You should use at least one hen and jake decoy. The jake is used to instill competition in the gobbler you are working.

Think of longbeards as whitetail bucks in rut. They compete with each other for available mates. Consequently, you should position your decoys approximately 25 yards from your calling position and place the hen and jake fairly close together. The jake should always be positioned where you intend to shoot because the longbeard will go to the jake to run him off.

Furthermore, visibility of your decoys is a problem all turkey hunters face, whether they are used in fields or woodlands. Early in spring the weeds and crops are low plus the leaves on brush and trees still haven't gone to leaf. That means visibility usually isn't a problem. However, as the season progresses, everything grows taller and gets greener as the leaves open, creating visibility problems. Therefore, it is important to have decoys like BuckWing's Lifelite Decoys that have mounting stakes that can be lengthened. This serves to get the decoys above obstructions so a gobbler can see them.

In addition, movement will bring your decoys to life. The BuckWing decoys move from left to right with the slightest breeze and this concept has been improved upon with the development of a new, patented Bobb'n Head Lifelite® Decoy. The whole decoy not only moves from left to right, but the head and neck move from side to side and up and down with the slightest breeze. This makes it look like it's actually alive and feeding. It will be introduced in early 2005.

With all the realistic improvements made in decoys, everyone needs to be aware of safety. You should put a blaze orange hat on before you attempt to pick up the decoys, especially in the woods. Without any blaze orange, another hunter sneaking in because he heard you calling may mistake your movement for that of a turkey, especially if the look-alike but "fake jake" is seen.

JIM FITSER
Allentown, PA

Jim Fitser is a retired school teacher who's hunted turkeys for 35 years. He's also an accomplished outdoor writer. His favorite call is the D-Boone De-ceiver made in 1970. Jim hunts Luzerne, Berks, Northampton and Pike counties.

FOGGY MORNING GOBBLERS

Any serious turkey hunter knows there is one constant about hunting gobblers— you NEVER stop learning. If you think you know it all, sell your Model 1100 and put your hiking boots in the closet.

Years ago I gave up trying to hunt gobblers in a steady rain. Until just recently I never bothered going if there was fog. Big mistake. But it took me years to learn that.

Hunting fog-ridden gobblers is best done in areas where you have access to a farm with woodlots on it, which you know holds birds. Be there at first safe light; often well after legal starting time on misty morns. Try one or two yelps, then listen carefully. Sometimes distant gobbles are harder to hear in the fog. If you hear a gobble, find a spot where you can set up in a hedge row between fields, with the gobbler in the woods on the far side of the field. Do not attempt to move into the woods where the bird is calling.

Set up and give a couple loud yelps. If he answers immediately, your chances are good; very good. Call back. If he again responds and appears to be coming in it's likely he's alone.

Between hearing his first gobble and your first reply, a hen and a jake decoy should have been placed in the field about 10 to 15 yards maximum from your set-up spot.

Note the direction of his gobbles as you probably won't be able to see him, even if he's close, depending on density of the fog. When he shuts up, swallow your diaphragm (not really) and stuff your slate or box call in your vest pocket. Keep watch for 180 degrees as, over the years I've learned, he will likely come in from one side rather than straight ahead.

Apprehensive? Insecure? Who knows?

But he will rarely gobble again.

There's something almost mystical about seeing a wary gobbler appear like an apparition from the fog. He may even appear and disappear several times before getting close enough to see the decoys.

Enjoy the experience and when he finally gets to the decoys, don't forget to shoot.

DENNY SNYDER
Lancaster, PA

Denny Snyder is host/producer of the U.S.A. Outback TV show and a free-lance videographer. He's on the pro-staff of BuckWing Products and Knight & Hale Game Calls. Denny hunts turkeys in Sullivan, Potter and Blair counties.

FOUL WEATHER MORNING?
LISTEN UP

You've been sitting in your truck dozing or watching raindrops splash on the windshield since 5:30 a.m. But now it's 8 a.m. and the fog still hangs heavy but the sun's beginning to peek through. Do you hunt as usual, find a known roost site, seek a field and put out your decoys or use some other technique?

Which option will you choose?

My first thought is to check my vest for binoculars because with fog still hovering on the hills it is critical to have good optics to scan the terrain for any birds that might have already flown down. I have also found that a good crow call after a hard rain is the perfect way to wake up the woods and entice that first gobble.

The morning rain creates a soft forest floor that will allow a hunter to quietly move in on any longbeard that responds to a locator call. The fog also provides the advantage of setting decoys in an open forest or out in a field, even in muted daylight.

Turkey hunting is a game of patience and if a hunter can sit still for an hour or two, the chances of lighting up and livening up as the morning progresses gets better. This will initially occur on the sunny side of the hills and fields where all the bugs and grasshoppers will be trying to dry their wings, making easy targets for hungry turkeys.

There is also the distinct possibility that the foul weather, if it was a storm, may have broken up a flock or two which will be trying to rejoin one another. Stay on the high ground for better visibility and the ability to hear those first regrouping calls or gobbles.

At such times I heavily rely on a Walker's Game Ear II to hear any activity in the area I am hunting. I do much more listening than calling, a facet of the hunt that many sportsmen ignore. Let the woods tell you what to do.

More than 30 years of turkey hunting has convinced me that being a good woodsman is more important than flawless calling ability.

CHAPTER 23

Uncooperative Toms

When Wary Longbeards Won't Answer

A thunder-clap reverberated from beyond the far mountain, doing what I'd failed to accomplish in nearly an hour of orchestrations. My array of instruments included two mouth calls, two box calls, a slate call, a tube call, a push-pin call, a crow call and an owl hooter.

My plan was to locate, then call to what I believed was the lone gobbler I'd located the previous evening when it sounded off a half-dozen times from its roost some 200 yards off the edge of a clearcut.

To my ear the calls sounded convincing. But somehow they failed to stir the tom.

That morning – a cloudy, breezy second-week-of-May day – I tested my entire arsenal of calls in hope of stirring the lusty tom. Finicky gobblers will often ignore certain calls, answering only the one or two appealing to them on a particular morning.

This dawn I'd have traded them all for a bass drum.

Hunting thunderbirds

Over the years I've learned that spring gobblers can't resist shouting back at thunder. The storm blowing in from the West, which at first seemed a bane, now became a blessing. Three successive claps in a five-minute period brought as many answers from the longbeard that surely heard my calls for the past hour.

No matter. Now I knew he was there. I also knew it was time for Plan B – which would require outsmarting the old boy by going easy on the chatter. When gobblers call but don't come, particularly in the first couple hours after dawn, it's a good bet they're in company of hens. Often they'll answer but not budge. Then it's a matter of simply waiting them out – or leaving for breakfast and coming back later in the day when their libido has been temporarily satisfied.

But rather than go free-lancing through the Pocono Mountains forest, I decided to cautiously close in on the bird, which was clearly hen-pecked for the time being, then set up and wait. The fat red tulip-poplar was a perfect backdrop. I settled in, pressed my back against the gnarled trunk, then offered a few soft yelps followed a minute later with some solitary clucks. That would be it. I'd follow the advice of old-time gobbler gunners whose techniques were to call once and hour, then sit back and wait.

The author heads from the woods with his "thunderbird" taken in the Poconos a few season's back.

To both my delight and chagrin the storm seemed to be moving away. While audible, the distant thunder was no longer loud enough to stir the gobbler. The flip side was that it was taking the rain with it and the sky hinted of pending sunshine.

It's not easy for a turkey hunter to resist calling, particularly when his pockets are jammed with all sorts sweet-talk devices. I pulled the Quaker Boy box call from my pocket, scratched fresh chalk on its edge, and set it beside me. Ditto for the Lohman double-slate.

Growing impatient, 30 minutes later I scraped out a series of soft yelps on the slate, then once again put it down. My plans were interrupted when another thunder-boomer shattered the air. Before the rumbles subsided the gobbler answered -- from nearly the same place I'd offered my renditions of a hen an hour earlier. Had I simply stayed where I was when I first heard the gobbler bellow, he'd probably have been sitting in my lap by now. Maybe hanging over my shoulder.

"O.K., you wanna make this hard," I silently challenged." Time to put Plan C into action."

Have a Back-Up plan

Every gobbler gunner knows there are no guarantees for outsmarting a bird – particularly a "thunder-bird" whose sole stimulation is a pending storm. Birds that return their messages reveal their positions. Birds that refuse to answer can be bumped at any time and a hunter might as well seek greener pastures when it happens.

But when things work out the result can be quite satisfying and you feel like a genius. My thought was that the bird had spent the early morning hours strutting and mating. Once his duties were done, he moved off to investigate the "hen" he'd heard along the edge of the field (Me). If I could lure him back to the route he'd taken to get to the field – one in which he was confident there was no danger – I'd have a chance of drawing him into range.

Ten minutes later I set up in a small clearing where fresh droppings and scratched leaves assured me birds would be by sooner or later. Again I offered a series of soft yelps, then adjusted my face mask, pressed against a stump and propped the gun on my knee, anticipating a long wait.

The black form moving through the greening woods several minutes later caught me off guard. The fact he was coming in silently was of no particular surprise but, considering his patience throughout the morning, I didn't expect him to appear as quickly.

At 90 yards, through the drab, shadowless woods, it was difficult to follow the bird as it unhurriedly worked through a stand of pines, then reappeared in an open plot of beech. Fearing the tom would bypass me I waited until he disappeared in a shallow bench, then cautiously drew the small push-pin call from my pocket, pressed the pin five times and peered down the ventilated rib.

The rest was anti-climatic. The unsuspecting gobbler poked its white head from the swale, then walked straight at me. I held the beads of my Remington 1100 SP on his neck for a full five minutes before he reached the magic 30-yard mark, at which time I triggered the load of 3-inch No. 5s.

It was a good tom, as Pike County turkeys go, weighing 19 pounds and sporting a 9-1/2-inch beard. Taking him wasn't one of those heart-pounding scenarios you'll see on a video nor did I stun them with my powerful repertoire of calls.

Simply, I outsmarted him. And I think about that hunt each time I look at the 8x10 hanging on my office wall.

Here in Pennsylvania – as well as other states with public hunting grounds – it's not unusual for bearded birds to develop lockjaw from time to time. After the first few days of the grand spring hunt birds have heard yelps and clucks from dozens of hunters. Chances are he'd also been spooked a time or two, making him as wary as long-tailed cat in a room full of rocking chairs.

Simply, educated birds require a switch in tactics. Don't be afraid to experiment. Even world-class callers with contest trophies lining their mantles won't bring a gun-shy bird into range.

Such has been the case time and again across the state as hunters here and there compare notes and share frustrations. They scratch their heads trying to figure out how to get the birds to answer their calls after the typical dawn flurry of gobbles – and take comfort in knowing that others are sharing the same frustrations. Sometimes it doesn't matter if you're a contest caller or someone who's afield the first time.

It's not the calling that turns the tables; it's a matter of good woodsmanship.

Stay in bed another couple hours

I asked several well-known Pennsylvania hunters how they adapt their techniques to outsmart tough turkeys.

"Last year I didn't even go out until 8 or 9 o'clock," said Arnie Hayden in my interview just prior to his death in 2001.

"Where I live a lot of the hunters who are out there long before sun-up go to work or head back home if they don't hear any birds on their roosts. Last year I shot my gobbler at 10:30."

Jody Hugill of State College, a member of the Lohman Game Call pro-staff and a popular seminar speaker on turkey hunting, agrees that timing is often the key to filling a tag.

"If I can't get a bird off the roost I'll just back off and wait until later in the morning," said Hugill. "By then they're likely to leave the hens and go looking for others."

He also recommends "easing up" when calling to birds made call-shy by hunting pressure, as I did on my "thunder-bird."

"You'll have to spend more time and be patient -- like going back to the old school of turkey hunting," advises Hugill.

Turkeys get tougher to hunt and less vocal as the season progresses.

"And don't get too fancy with the calling out of desperation. A few soft yelps or clucks, that's all you need. The gobblers may not answer but they know you're there and will often come in later to look for you."

Joe Hall, owner of Hally's Callers in Doylestown, Bucks County, has hunted gobblers from his Potter County cabin for more than 25 years and has enjoyed success by traveling off the beaten path when toms become call-shy.

"You'd be amazed how many hunters go out before the season to try to videotape turkeys," said Hall. "In the north lots of hunters scout and call up birds in March and April. Turkeys are educated long before the season starts."

That, combined with early season hunting pressure, puts the clamp on many toms.

Hall said he "travels the extra mile" when hunting public lands, locating gobblers that haven't experienced as much pressure and don't frequent areas near roads, where hunters often cruise and call.

"Today's turkey is a whole lot tougher to hunt than those of the early 1970s," Hall believes. "They're educated birds and the hunter has to put forth some extra effort to get them."

Of course, not all of Pennsylvania's turkeys are shy and discreet. Over the past 25 years my wife and I have taken longbeards only minutes off their roost trees. By 8 we're back at the truck sipping coffee. Or maybe we've located gobblers that simply couldn't hold their tongues, bellowing back at every call with a rare few actually coming in on the run.

But when the season's more than a week old, wise gobblers instinctively know there's something amiss in their woodland haunts and many simply clam up.

That's when woodsmanship, craftiness and timing pay off, even if you're not a champion caller.

STEVE LECORCHICK
Northern Cambria, PA
Steve Lecorchick is the promotional
manager for Penn's Woods Calls. He
has won or placed in more than 100
calling contests. Steve and his partners
ply the woods of Indiana, Cambria and
Clearfield counties.

DOUBLE-TEAMING
STUBBORN GOBBLERS

As a professional guide who hunts for a goodly portion of my living, I seldom get
to hunt alone. I have used several different tactics on many different gobblers over
the 30- plus years that my fellow hunting partners and I have been chasing them.

The most common scenario is to sit my hunting partner right next to me. He
or she is usually the shooter and I am the caller. I prefer this set-up the best for
a couple of reasons. First, most of the time the gobbler will come to the calling
position and, second, it is very helpful to be able to communicate with one another
as the hunt is in progress. This is especially important if you are hunting with a less
experienced hunter. In this close proximity I can offer advice such as where the bird
is situated, how far away he is, where to point the gun and, most important, to keep
the hunter still and help judge the distance.

Another technique I like to use is to start the hunt by sitting next to the shooter
where I will call for a short time. Then I'll move straight back about 25 yards
behind the shooter and call from there. If all goes well and the birds keep coming,
I'll stay put. If the tom doesn't seem to be moving any closer I will move back even
farther, calling as I retreat another 25-30 yards.

By doing this, I'm hoping to make the bird believe the hen that was once
interested is now disinterested and leaving the scene. After I call a short time I will
become quiet and wait him out, all the while being very alert as there's a good
chance he'll come n quietly. If he continues to gobble but doesn't come our way I
will call softly with a few clucks and low soft yelps.

Another method is to sit about 10 or 15 yards from my partner and both of us
will call, simulating a couple of different hens. Any two calls that have different
sounds will do. When using this method, be sure not to get in a calling duel
with each other.

No matter which double-teaming methods you use, remember the most
important consideration is to be safe and enjoy each other's company.

CHAPTER 24

HUNTING EDUCATED GOBBLERS?

Slow Down and Shift Gears in Late Season

Eddie Salter gave it everything he had; mouth calls, box calls, slate calls and aluminum calls were all brought into the standoff.

But in two hours of yelping, cutting, putting, cackling, purring and gobbling, then walking away and calling from a distance, shifting from one call to another and even duplicating a gobbler's "spitting" and drumming, we couldn't lure the bird close enough for a shot. Even the silent treatment failed to bring the stubborn tom into sight.

This Tom (me), holding a Beretta full-choke shotgun loaded with Winchester number fives, kept a watchful eye on the field to the right and the dense woods out ahead. Although our intended prey gobbled with regularity from 40-50 yards off,

Eddie Salter, an occasional Pennsylvania visitor, is one of the nation's best-known turkey hunters.

he wasn't about to make much of a move. It was the classic standoff and I relished the moment, enjoying the live seminar put on by Salter, one of the country's best known turkey callers.

The glitch, as we later established, may have been our red pickup truck parked at the end of the field where we first heard a loud gobble and quickly moved to set up. Previously we'd stopped at other promising spots in hopes of getting an answer via "road calling." Finally, on Salter's seventh or eighth try his yelps drew an immediate answer from a bird no more than 100 yards off.

We abandoned the truck where we'd stopped, then cautiously worked to the edge of the woods bordering a five-acre field. Thirty yards inside the woods we found separate trees and began what would turn into a marathon, of sorts. There we spent more than two hours calling and watching. From occasional sightings through the leafy woods we noted an occasional hen scratching leaves; one wandering within 15 yards a couple times.

The gobbler eventually strutted into shooting distance but through the limb- and deadfall-littered understory I hesitated to shoot, even though I could readily see the bird's red-and-white head and the arc of his fanned tail no more than 30 yards off. But I didn't yield to the temptation.

Finally, at noon, the bird simply walked the opposite direction, leaving with a final gobble that sounded something like: "Tom Turkey one, Tom Fegely zero," as we scratched our heads and admitted we'd been beaten.

The encounter took place in the rolling woods and farmlands of northwest Iowa where Judd Cooney, a well-known outdoor writer, runs a trophy whitetail and turkey hunting camp. His dozen-plus leases also hold wild turkeys; big wild turkeys weighing as much as 25-28 pounds.

Salter, a regular visitor to the Keystone State for seminars and hunting, is one of the nation's most respected and knowledgeable callers with countless trophies and plaques covering his walls. He's hunted dozens of states in his advisory and public relations position with Iowa-based Hunter's Specialties, makers of game calls and other products of interest to hunters. He's also been the subject of numerous deer and turkey hunting videos.

Our lengthy encounter occurred on the final morning of the Iowa hunt which, said Salter on the way to our morning hunting grounds, would demand some new techniques and tactics. By now the gobblers had heard and probably seen it all. Toms that were hunted and survived had toned things down and were neither as reckless nor lusty as they were a few weeks earlier. Now, with hunt's end a couple days off, yet another educated tom would make it through the season.

Knowing I'd be returning to Pennsylvania for our gobbler season, I asked Salter for some advice to share with Pennsylvania hunters whose tags were still intact.

Bearded Delights and Satellites

"At the start of the season you'll have dominant gobblers with hens and satellite

It is sometimes possible to tell whether a dropping was left by a hen or tom.
Gobblers leave J-shaped scat and hens make rounded droppings.

gobblers moving around a lot trying to find a hen here and there," said Salter. "It's those satellite gobblers that are more likely to be called in and shot the first couple weeks of the season."

"As season's end approaches the old gobblers will still have the locations of some hens pegged," he continued. "And they're not going to be very willing to leave them."

Satellite gobblers, so named because they circle the fringes of older birds' routes, are usually two-year-old birds who may have been whipped by a boss gobbler or two several weeks earlier.

One trick Salter uses – a standard technique for fall hunting but seldom used in spring – is to break up the flocks as the birds get ready to roost. He suggests waiting until about 10-15 minutes before fly-up to bust birds he's located. With dark approaching, they'll then roost in scattered sites and, at dawn, fly down and seek to join ranks. Being there early and setting up among them and calling – similar to the fall ploy – often brings results.

Switch to deer hunting tactics

Shifting into low gear for the final week or so of the season is Salter's most important bit of advice.

Said Eddie: "When I hunt any state on the first days of the season I like to be very aggressive but as the season moves along I change to deer hunting tactics."

Turkey sign such as droppings, scratchings, tracks on backwoods roads and dusting sites are noted, just as a buck hunter would look for tracks, rubs and scrapes. Salter sets up in places birds pass through each day, as substantiated by the field sign. At such sites he issues non-aggressive calls.

If a gobbler answers, he gets as close as possible, then sits down and tests his patience. Often, heavily-hunted toms will come toward the call but won't necessarily answer after a "contact" gobble or two. If he has hens, a wise gobbler will let them do the dirty work and move toward the yelps, clucks and cutts first. If they're not spooked, he may follow.

On those frustrating occasions when you know a gobbler's there but he won't close those last critical yards – as occurred on our Iowa hunt – Salter will carefully retreat, then swing around to the place he first heard the bird gobble.

"In that way turkeys are like deer," said Salter. "Just like a buck, an educated gobbler will often turn around and travel the very same route he came from. He probably figures he was safe the first time and will be safe again."

Salter said many hunters, out of due respect for a turkey's hearing and eyesight, will not change locations for fear of spooking the bird or its hens. But making a wide arc and setting up where the birds may have been feeding, preening or resting originally is one of the most effective methods for luring in late season birds.

Locate lone hens

Observant hunters will also note areas in which lone hens are seen, particularly on the edges of fields, overgrown orchards, woods with heavy understories and other sites where they nest. Most hunters believe that nesting hens will not join up with gobblers but that's not true, Salter's learned. As the season progresses, hens break off to lay eggs – one per day – but don't begin full-time incubation until full clutches of 10-12 or more eggs are laid. Until hens begin serious and regular incubation, they'll continue to travel in small flocks, often with a gobbler or two.

Gobblers find those hens and each day they stop by nesting areas, knowing that a hen or two is about. Setting up on the edges of overgrown fields or in moderate understory, perhaps with a decoy or two as extra enticement, often proves effective.

Again, patience and perseverance are the keys.

Eddie's also learned that loud, aggressive, back-and-forth chatter between gobbler and hunter is more an early season tactic. Later, when gobblers aren't as likely to rattle the woods, Salter offers more subdued calls.

"When a gobbler gets close you don't want to sound like a 1,000-pound hen," he laughs.

"Any bird that survives through the last week or so of the season is going to know something's wrong but he'll leave evidence that he was there. That's good enough reason to slow down and become more of a deer hunter."

Lone hens off the nest and feeding are attractants for wandering gobblers.

CHARLIE BURCHFIELD
DuBois, PA

Charlie Burchfield hosts Gateway
Outdoors, a Northcentral region
radio show, and is a field editor for
Pennsylvania Outdoor Times. He
hunts in Clearfield, Elk, Potter and
Blair counties.

YOUR GOBBLER WANDERED OFF? CIRCLE HIM

The bird you've been calling for two hours decides he's had enough and wanders off.

Now what?

Most likely the gobbler has hens which dictate his travels. If the calling duel took place during the first half of the morning and you have time, allow the hens to wander off to their nests and patiently await his return.

Generally, when a calling situation like this takes place, Mr. Tom, or another gobbler traveling with the group will return and attempt to "pick up" the hen (me) left behind. However, there are other options.

After the direction a gobbler travels away from your calling location is determined, try the "cutt and run". Give the bird(s) a wide birth and parallel his path of travel. Move with caution as there may be other hunters in the area who have heard the bird as well.

Also be aware of breaks in the terrain or other features such as logging roads, pipelines and fields that may adjust the bird's travel plans or attract the birds. This is where knowing the lay of the land is important.

Don't overcall. Instead use locator calls to entice the bird to gobble as a means of keeping track his travel. Your chances of tagging the bird increase when you've traveled far enough to set up; hopefully far in front of him.

Keep in mind that when hunting turkeys there is no hard and fast rule that works every time. However, as you gain experience in the turkey woods you'll learn to read the situations that recur. Use the knowledge to get Mr. Tom strutting toward the end of your shotgun.

If it is possible, preferably before the season, spend time in the area and devote a morning or two to simply listening. Follow the birds and learn their daily routine but don't disturb them by calling. If you do, by opening day they may simply ignore your calls.

BEN MOYER
Farmington, PA

Ben Moyer is a familiar name to hunters
and anglers as a columnist for the Pittsburgh
Post-Gazette. He's the author of Out
Back: Reflections From the Appalachian
Outdoors. Ben is a member of the
Pennsylvania Turkey Hunters Hall of Fame.

EVEN QUIET TURKEYS LEAVE FRESH SIGN

I think hunters overlook the importance of sign, especially in the spring when most
are focused on hearing gobbles. If you hunt near fresh sign – scratchings, droppings or
fresh tracks in mud, dust or snow – you can have confidence that turkeys are nearby.

Fresh sign is particularly important in Pennsylvania's highly pressured turkey
woods. Gobblers do not seem to be as vocal as in years past, but they still scratch,
travel and leave droppings as they always have. Fresh sign gives you a clue as to where
the gobbler that ignores your calls disappears with his hens after they leave the roost.

Rarely can a caller turn that gobbler away from his path, but if you set up in their
feeding or strutting zone before their arrival, you don't need to call the bird so far.

I hunt better when I keep my ears tuned for any turkey sound, not just gobbles. If
you're not selectively listening for gobbles only, you learn to hear faint clucks, birds
flying up to roost in distant hollows, or scratching in the leaves just over the ridge.
Any time you know a turkey's whereabouts you have an advantage.

Learning to listen is a huge part of turkey hunting, and gobbling isn't the only
sound that can tip you off. Here's an excerpt from my book of reflections about
hunting and fishing – 'Out Back; Reflections From the Appalachian Outdoors" – that
tries to capture some of the sounds of the turkey woods:

"In a dry October you can walk up close to turkeys on days when the leaves rattle
underfoot like hail on a barn roof. That is, if your pace is erratic and unpredictable. To
explain this, reverse the roles. Listen to a flock scratching in dry October duff. The sound
will fill your head and chest, but unless you can see the birds you will turn your ears in
every direction, straining your senses, triangulating off the tree trunks, searching, but you
will not be able to pin down the direction of its source.

I have watched turkeys pitch up into grapevines to feed on the musty fruit. They
teetered there, dipping forward and back, clutching at vines and flailing the thicket with
their wings for equilibrium, sending showers of broken tendrils to earth in a racket that
scared deer off the ridge."

CHAPTER 25

Do You Hear Me Now?
Respecting a Turkey's Hearing Ability

Several years ago I took a friend – let's call him Jeremy to protect the guilty – on his initial spring gobbler hunt. When we got within about 100 yards of a gobbling bird I set him up 25 yards in front of me and told him which way to face.

I called softly, stroking a slate and offering subdued clucks, purrs and super-soft yelps. Then I turned the opposite direction and did the same thing, hoping that the tom would think I was a hen moving away. For the next 10 minutes all was quiet. Then I heard the unmistakable crunch of dry leaves and hoped that Jeremy was also hearing them.

He didn't.

When the bird showed up it caught him by surprise. When he lifted his gun, the longbeard took wing.

"All of a sudden he was right there," Jeremy mumbled a minute later, as if about to break down and cry. "I never saw him coming but all of a sudden there he was."

"How come you didn't call?" he challenged, as if it was my fault he'd blown his opportunity.

"I did call," I insisted.

"I called that bird in. He didn't gobble but I heard leaves crunching 100 feet away. That's when I purred. Didn't you hear that?"

"No," he admitted. "I never heard you call or the turkey coming."

Although Jeremy couldn't hear the subdued yelps and clucks, the gobbler picked them up at, I suspect, 75 yards, attesting to his acute hearing ability.

My low-volume calling did the trick. Had we not received an answer in five minutes or so I planned to boost the intensity a bit, but this time it wasn't necessary.

Over time, similar happenings have convinced me that hunters don't give due respect to a turkey's hearing abilities. When you have the opportunity, watch a hen or gobbler that has no idea you're there. You'll see it looking and listening as often, or more, than it's feeding.

Tread softly and safely

Many hunters will unwittingly spook unseen birds by making too much noise,

Late season hunting means taking advantage of the heavy vegetation and moving slowly and quietly to find birds, all the while wearing an orange hat.

then wonder why the gobbler didn't come. Then again, the hunter probably didn't know a turkey was there in the first place and may have been careless in approaching and setting up. The give-away may have been calls rattling around in a pocket, an errant sneeze or cough, a twig snapping, crunching leaves, your back rubbing against a rough-barked tree, rocks grinding as you walk on them or opening a Velcro-closured pocket.

I've also watched as a partner flicked his safety and a gobbler at 25 yards froze for a mini-second, then sprinted off through the brush. Any one of the indiscretions can be enough to put a bird on alert.

Turkeys are accustomed to hearing sounds of all sorts, from traffic on a highway or a plane flying overhead to dogs barking, cows mooing and crows complaining. But that doesn't bother them. It's a daily occurrence. Let them hear a strange sound, however, and they'll immediately be on their toes.

To this day my friend still doesn't believe that I called that gobbler to us on that warm May morning. I suggested he invest in a Walker's Game Ear before his next time out.

Do You Hear Me Now?

Turkeys have the unique ability to hone in on sounds; the downfall of more than one hunter's quick set-up and some errant noise that pinpoints the source. Even though turkeys possess no "flaps" on their ears to concentrate sound waves, they have an acute ability to hear things most of us don't. In researching this book I could find no reference to any studies that have been done on the hearing abilities of toms and hens. It has been suggested via field observations, however, that turkeys best hear low-frequency sounds and can tune in long distance sounds.

It stands to reason that both abilities are particularly useful during the spring strutting season when gobblers drum consistently and the unusual sound is picked up by other turkeys. I've hunted with men who could hear

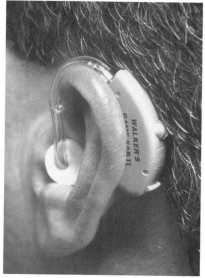

The Walker's Game Ear can help hunters pick up far off turkey talk and footsteps in dry leaves.

the bass sound from more than 100 yards off on quiet days or when a slight breeze carried the drumming frequency. However, I can hear the drumming at only 15-20 yards or so, hindered by my inability to hear low frequencies.

A turkey's low drumming sound is similar to that of the ruffed grouse in that a listener has to know what it sounds like before it can be heard.

No idle chatter

My friend Will Primos of Primos Wild Game Calls always demands that he and a hunting partner remain at least 10-15 yards away from one another when walking and calling to locate birds. This not only discourages idle chatter but also makes it easier to detect distant turkey sounds.

I ask the same thing when with a partner. On more than one occasion I, as caller, didn't hear the gobble of a bird because it responded before I'd completed the call. But he, standing a dozen or so yards away, picked up on it.

Quiet hunters and noisy turkeys are the keys for hearing and revealing the location of the hens and toms that make our days. Although the phenomenal vision of turkeys is widely known, hunters will do well to pay more respect to the hearing abilities of our feathered quarry.

Keep all this in mind when you're calling, scratching leaves to simulate turkeys feeding or simply moving from place to place. And if you're hunting with someone else, remind one another to be as quiet as possible, keeping chatter to a minimum.

Often we will fail to recognize our own indiscretions while quickly noting someone else's.

FRANK ANTONIACCI
Moosic, PA

Frank Antoniacci is president of the
Lake Region Longbeards. He carries his
Remington 870 on gobbler hunts in
Wayne, Pike, Monroe and Lackawanna
counties and into northwestern N.J.

USE THE WIND
TO YOUR ADVANTAGE

Hunting gobblers in the wind can be a very difficult task. Although your hunt
becomes less than ideal, numerous techniques exist to handle the cards dealt by
Mother Nature.

Here are some helpful techniques to be tried on windy days:

- Before you head out, check the local radio report for the day's forecast.
- Use calls that are loud and carry long distances.
- When you're calling to locate a turkey, pause until breaks in the wind ensure the
  sound of your call isn't diluted.
- Set up in familiar areas and call as often as you think is necessary. In the early
  morning, listen in open areas, not close to background noises of leaves and
  branches blowing about and creating interference.
- Look for other clues as to the presence of turkeys such as tracks, feathers,
  droppings, scratchings and dusting bowls.
- Utilize geographical boundaries such as low spots in the terrain, creek bottoms,
  valleys, and the backsides of hills and mountains where the wind is less prevalent.
  Fields and open areas are also very productive.
- Use camo to your advantage. Windy days permit the hunter to post in more open
  areas due to the constant movement of surroundings.
- Be aware that toms answering you or gobbling on their own will often sound
  farther away than their actual location.
- Realistic looking decoys are an excellent aid on windy days as they will move,
  imparting a lively look. Be sure to use quality decoys. Turkeys may shy away from
  decoys that don't look like the "real McCoy".
- Finally, windy days typically allow for more hunter movement. If you are
  fortunate to have a gobbler come into range, don't be afraid to cautiously and
  slowly move your firearm and properly position your sights on the bird. Keep any
  such movement close to your chest and move when the wind is at its peak.

HARRY BOYER
Newport, PA

Harry Boyer prefers the "Classy Hen" call, one of several that he makes and sells. He's a popular speaker on turkey calling and hunting at sportsmen's clubs and with youth groups. Harry sticks to Perry and Juniata counties for his spring and fall turkey hunts, often taking others afield with him.

LET THE TURKEYS HUNT YOU

Turkey hunting to me is hunting them until they find you.

This may sound backwards but it is exactly what happens. I walk and call and listen for an answer. Once I hear another turkey answer my call, I get into position and wait for it to find me. Often they come in silent and unseen. That's when some wise decision making comes into play.

For example, knowing when to pull your shotgun to your shoulder is a requirement. Since turkeys can see almost 300 degrees at any given time, you need to keep still until they get their heads behind trees or some other object that will shield you.

In some cases it is imperative to let a gobbler walk away even though he may be right in front of you and as close as 10 yards. It's made even more difficult when other gobblers are with the tom you are hunting, which means other eyes are watching your every move. Hens with a gobbler will give your position away in a mini-second if you move and they see you.

In situations like this I prefer to let the gobblers walk away, then try to call them back. I use a hen call I've heard many times which sounds like "quot-quot," offered without any pause between the two sounds. Many times this call alone will bring a tom running back to your position.

Another important factor in successfully hunting spring gobblers is to try to detect the gobbler's mood. I've noted in my 36 years of turkey hunting that the mood and emotional state of a turkey can change at a moment's notice. One minute he can be quiet as a mouse and the next he starts gobbling his head off. Adjust your calling according to his mood.

The three most important things a new turkey hunter must learn is: Don't move, don't move and don't move.

That's especially true when a turkey's hunting you.

CHAPTER 26

DANCES WITH JAKES

Sometimes It's the "Shortbeards" That Make Our Day

Jakes typically travel with their own kind, from two or three partners to as many as 10 or 12.

I think of jakes as the spike bucks of the turkey world. For many years, up until the antler restrictions on Pennsylvania bucks, spikes were legal game for hunters. Now they may be legally taken only by junior hunters.

Not so for the short-bearded residents of the turkey woods which remain legal game for everyone.

But they do present a dilemma for some hunters: To shoot or not to shoot?

Personally, I have no problems with anyone taking a jake – first-timer or old pro and anyone in between. It's largely a personal decision and one that should be given some thought before going afield.

Jakes are the teen-age residents of the turkey world and inarguably are easier to call and kill than their 2-, 3- or 4 year old counterparts. I've called to jakes that practically ran down my barrel and others that milled around me for 15-20 minutes, paying little but casual interest in my camouflaged form pressed against a fat tree.

In short, they simply lack the experience necessary for making wise decisions similar to human teenagers who must test their own immortality or "wisdom" from time to time and often pay dearly for the education. For turkeys, things are much the same as they move into their first spring. Their initial lessons occur when their mystical spring lust drives them to join up with older birds that may have condoned their companionship for a time in fall and winter. By February and early March, when the initial gobbles echo through the leafless countryside, jakes slowly get the hint that, as far as longbeards are concerned, they're second-class citizens. But they're not sure why.

Trouble-prone youngsters

Like teenage Homo sapiens, young Meleagris gallopavo silvestris nomads "just don't know how much they don't know," a turkey hunting friend of mine offered one morning when we came across a gobbler chasing three jakes across a harvested bean field. One bird stopped to watch his cronies who were behind him when another gobbler sprinted from the nearby woods and attacked the curious jake, flailing his legs in hopes of getting in a stab or two with his spurs, then grabbing the young bird by its neck and hauling him around in a circle several times. The youngster got the message and quickly flew off.

Jakes not only become increasingly wary with each such incident but often become close-mouthed and nervous upon hearing calls produced by longbeards and hunters alike. The lessons continue throughout the mating season as they travel in small flocks of three or four each. Although they may continue to respond to nature's urge to mate for their first time, they become more cautious in approaching a flock of hens, especially when a longbeard may be tending to the harem. Taking

An angry gobbler shows his discontent with a jake as a short-bearded audience looks on.

advantage of safety in numbers, however, jakes have been known to gang up on bigger toms from time to time.

This isn't to say that educated jakes are pushovers. Although they remain susceptible to calling, they often can't resist a decoy set-up or the sight of a half dozen hens feeding in a grassy pasture. Although one of the first lessons jakes seem to break is that strutting is "verboten," a more dominant shortbeard will sometimes strut in its attempts to interest a hen but will quickly fold its feathers if a longbeard's heard or seen.

Jake gobblers will gobble

Jakes will also gobble, particularly on the roost. The wiser birds, however, will keep their gobbling to a minimum when on the ground as it will raise the ire of longbeards within hearing range. After hearing it a time or two, a hunter soon learns to tell a juvenile's gobble from that of a mature tom. The sound is best described as a broken and somewhat incomplete and disjointed version of the adult's resounding gobble-obble-obble. That said, every hunter gets fooled sooner or later, especially when a trio of jakes somewhere along a ridge all offer their love songs at the same time.

Depending upon how much understory a pair of gobblers must pass through as they approach your woodland set-up site, it is often difficult to determine whether a longbeard or shortbeard is in your sights until they close the distance. A binocular will aid in making a far-off I.D.

A jake's fan shows that several of its central tailfeathers are obviously longer than the others.

The most convincing clue as to a juvenile's identity is its spread tailfeathers. A strutting jake will show about 18 tailfeathers with 3-4 central tailfeathers longer than the others. When the tail is folded, however, the longer feathers do not show any distinction from the others.

A year-old jake's beard grows to about 3-5 inches by spring although some late-hatched birds may show no beard at all, except a nub lost in the breast feathers. "Look for the beard" is the cry of safety-conscious spring hunters. Jakes weigh about 15 pounds, give or take, and appear much slimmer than mature toms. They also hold half-inch spurs.

Although it's been a long time since I shot a jake, they have played varied roles in my search for and success on longbeards. Few hunters are at a loss when it comes to recalling times when jakes– although not shot at – have played a part in a hunt. One such memorable encounter occurred just over the Pennsylvania border north of Bradford County several years back. It demonstrates how curious jakes that you have no intention of shooting can, nevertheless, influence and add a touch of humor to your hunt.

Dances With Jakes

The morning's hunt was dwindling to the final, precious hour as the sun yielded to a raw, breezy, cloudy front pushing in from the West.

We'd spoken briefly with a lone gobbler four hours earlier but once off the tree he'd obviously joined his harem, thoughtfully announcing his route in the opposite direction from our hillside setup until only a feeble gobble could be heard.

Following a 30 mile round-trip to other turkey haunts in New York's southern tier mountains and farmfields with little more than calls from woodpeckers and the bellowing of cows to break the monotony, we decided to head back to camp and compare notes with others out to fill their tags.

Let's give that hill a try again," I offered as we approached the spot where we'd spent the opening hour. "Maybe he's done with his hens and out shopping."

My companions Paul Butski, Bob Walker and Wayne Davenport (who had arranged for our hunt with area landowners) agreed. Five minutes later we were parked next to the old home place where we'd impatiently awaited the dawn.

Butski, a well-known call manufacturer and turkey hunter, Walker, inventor of the popular Walker's Game Ear, and I were in the Candor area of southcentral New York for a National Wild Turkey Federation celebrity hunt at nearby Turkey Trot Acres. Localite Davenport, intimately familiar with the hilly, farmland region west of Binghamton, offered to serve as our silent guide.

Four-on-one isn't the accepted method for pursuing wary gobblers but all of us have logged many years of turkey quests. Then, too, we'd agreed that Butski and I would be the "participants." Walker had bagged a big gobbler a day earlier and Davenport wasn't carrying a gun. Nor was Butski, who gets a bigger kick out of calling in birds for others than shooting them himself.

Jakes will sometimes strut but not when adult longbeards are within sight.

A second chance

When Paul cupped his hand and broadcast a series of high-pitched yelps and cackles we tilted our heads against the wind, listening for a response. It came within seconds later and within the minute we'd pulled on gloves and masks. I loaded my Remington 1100SP and told Walker to do the same.

"Maybe I'll need backup," I chuckled, as he dropped shells into his gun.

Within the minute we were sprinting across a stream and up the far hill, hoping to intercept the turkey which we agreed was some 300 yards off.

"We'll stay out of your way," Walker assured as we paused to catch our breaths. "He's all yours, Tom. I'm not gonna shoot."

As we drew near, Paul again yelped, this time drawing response from inside 100 yards.

"Right here," Paul ordered as we both scrambled to break off small shoots growing against the tree where I wanted to sit. In my haste I poked him in the eye with the butt of my gun although I didn't notice my indiscretion at the time. As promised, Bob and Wayne disappeared to the spectators' seats somewhere behind us.

By now the gobbler had moved closer although he was not yet visible in the low undergrowth upon which I steadily held the sights. A series of subdued yelps and cackles was immediately followed by a double gobble, bringing that pleasant feeling that this bird was "in the bag."

But having many years of gobbler chases under our cumulative belts, we knew better than to jump the gun.

With the Remington resting atop my knee, I peered down the vented rib and aligned the glowing red and green fiber-optic sights. I cautiously shifted my gaze, noting a sparsely treed opening atop the shallow rise to my left. If the bird hesitated in coming through the scrub growth he'd likely feel more confident moving into the open woods where more landscape could be viewed. There he could get a better look at the hen he suspected was drawing near. I'd need to swing and fire quickly, however, as his silent appearance would surely come fast.

Then again, he could simply pop out of the greening understory and give me a 25 yard shot. Several long minutes passed without a sound as I continued to anticipate the sight of a red head poking through the scrub at any moment. Without moving my head, my eyes regularly flicked left, as far as my peripheral vision permitted.

Butski called softly but all remained eerily quiet. A full 3-4 minutes of penetrating silence followed, bringing the certain feeling that something was about to bust loose.

And it did.

A jake's beard grows 3-5 inches in its first year.

Dances with jakes

A loud, guttural, alarm-laden yelp off to my rear left was followed by the shufflings of what seemed to be a dozen turkeys scrambling for cover. Seconds later a dark form appeared in the shadowless woods, making a beeline for cover.

Confused, I held my fire. But Walker's aim was true as he dropped the sprinting gobbler at 15 yards. Hardly had the echo returned from the far hill when Davenport appeared, diving on the thrashing gobbler and pinning it to the ground.

Bewildered, I lowered my gun, unaware of what had taken place. It wasn't until things calmed a bit that Walker explained what had occurred; all within his view but out of mine and Butski's.

As Bob awaited the report of

my gun while perched on a deadfall 30 yards behind me, he heard a shuffle in the leaves behind him. The Game Ear in his left ear had amplified the sounds of crackling leaves, which were being made by four jakes coming in to Butski's calls. He and Davenport froze as the jakes trotted by at an acorn's throw.

"At first I thought it was a couple deer coming," Walker explained. "I didn't move, then I saw the jakes going by."

As anticipated, the gobbler had attempted a circular sneak but, probably hearing then seeing the quartet of short-bearded intruders, he threw caution to the breezes. The shuffling of leaves I heard was the sound of the longbeard chasing off the young gobblers. They'd expected to find a lonesome hen but instead ran into "the boss."

The lead jake, surprised by the bigger tom's charge, putted loudly, kicking off the confusing episode. The gobbler which had circled Butski and me sprinted by Walker and Davenport at only 10 yards. As the big tom headed toward the confused jakes, Walker lifted his gun. Correctly assessing the confusion and assuming I had no shot, he tripped the trigger when the tom put a bit of distance between itself and a lingering jake.

Later, Walker also tripped across his tongue apologizing; feeling guilty at having taken "my" bird. Butski and I poured more fuel on the fire, playing the role for what it was worth.

"Oh sure," I offered without a smile. "You tell me "It's all yours, Tom, I already shot a bird, then as soon as it shows up you shoot it out from in front of me."

"What a friend you are," I muttered, turning and walking off to retrieve my shotgun. "It's all mine, alright."

I couldn't push things any further without laughing, as Walker again told what happened.

Although I didn't dare admit it at the time I was glad for Walker's "back-up" but we all enjoyed rubbing it in that he shot MY gobbler. Bob continued to apologize all the way back to the truck with Butski, winking at me a time or two despite a golf ball size shiner next to his nose.

Making the record book

Then came the clincher. The trophy longbeard not only won the weekend's award for the biggest bird in camp, but also qualified for entry into the NWTF's record book. Back at Turkey Trot Acres, host Pete Clare weighed and measured the big gobbler which scaled out at 22 pounds. Its beard taped 11 inches and each foot held 1-5/8th inch spurs.

Sitting back with a brew that evening while telling turkey tales and congratulating Bob around the fireplace, it struck me that those lovable, undisciplined and comical jakes had played a major role in the morning's never-to-be-forgotten hunt.

Thank God for the shortbeards.

BARRY HAYDT
Danielsville, PA

Barry Haydt guides hunters spring and fall in eastern Pennsylvania. He spends most of his quality turkey time near home in Northampton County and in Monroe, Pike and Carbon counties in the Poconos.

JAKE OR LONGBEARD?
MAKE A QUICK I.D.

A red-headed turkey has been spotted through the green woods and you get your shotgun in position for the pending action.

As the trio of birds gets closer you start looking for a beard. But the understory prevents you from making a decision. Then one of the jakes breaks stride and fans his tail for a moment. That's when you note he is not a longbeard. He's a yearling tom – as are his traveling companions – and you allow them to come close, just for the fun of it.

How do you quickly tell if it's a jake or not? Fanning will help you decide. One sure way of telling a strutting mature gobbler from a jake is that the full-fanned tail of a mature gobbler is round. A jake fan has the center tailfeathers about 3-4 inches longer than the rest of the fan.

The second clue, which you eventually notice, is a short beard measuring 2-4 inches. Often it's held erect for a short time when the bird becomes excited. However, some jakes have no detectable beard and that makes them illegal to shoot.

You will also notice that jakes' heads look like mature toms' heads, but are not nearly as brilliant. A jake is 10 to 12 months old by the time he reaches his first spring season. He's the "teenager" of the turkey world and, like human kids, he'll often do some dumb things. He may be lonely, anxious and frustrated in his first breeding season and this makes him easier to hunt.

I've found that jakes respond well to low pitched yelps of other jakes and the raspy call of an older hen. The gobble of a jake is usually not as full and robust as a mature gobbler and can be identified by a hunter. But if a group of jakes answers your call they often do so in unison. From a distance, a trio sounds like the best gobbler on the mountain. Even the most experienced of turkey hunters may have a hard time telling the sound from that of an old-timer.

If you find jakes coming to your decoys but you decide not to shoot, sit back, stay still and enjoy the show. It's all part of this grand, spring sport.

ROGER HAYSLIP
Quakertown, PA

Roger Hayslip is a youth minister and a newcomer
to the outdoor writing field. He's called in
numerous longbeards and jakes for himself and
his young companions in Bucks, Lehigh and Pike
counties. He's also president of southeastern
Pennsylvania's Quality Deer Management chapter.

WANT TO SHOOT A JAKE?
NO PROBLEM HERE

"Wow, that's a big bird you've got there," said the grizzled old veteran in front of
the whole crew. Twelve-year-old Andrew Hargrove's eyes danced with excitement
and a broad grin arched across his face. It might as well have been a 40 pound
gobbler, with a 20 inch beard and four inch spurs. Never mind that there was barely
enough short, protruding bristle to clearly see. This was a jake and it was all his.

Give the little fellow his due respect – the bird that is, not the kid. Jakes
have long been the youngster's or novice's best friend. But what serious turkey
hunter among us has honestly never in his life strained his or her eyes for any wee
little sign that a jake had the word "legal" written on it? A plump and juicy jake has
bailed many a hunter out of a dry and empty season (and they're much better to eat
than tough, old toms anyway).

That said, I hunt for the big, bad boys as long and hard as all the rest.
Who wouldn't prefer a nice gobbler, given a choice? I seriously doubt that there is
a hunter anywhere on the planet that would bowl over a jake standing side by side
with a mature gobbler, and let the big one walk.

Being an archery enthusiast with about equal interest in Ol' Mossy Horns and
The Bearded One, I do all of my fall turkey hunting simultaneous with the pursuit
of rutting deer, using only a bow and arrow.

Under this particular set of circumstances, I actually sometimes prefer a jake or
hen during the fall season. I would rather leave Ol' Mr. Longbeard in place for the
spring dance. Besides, the successful taking of any turkey by archery methods will
make even a jake suddenly appear like Godzilla. While we're on the subject of jakes;
if you don't have a jake decoy for the spring season, run (That's right, run!) to your
nearest store and get one. Years ago I read an article that said in a spring calling
situation a gobbler will approach the jake decoy over the hen every time. I have
thoroughly tested that theory for several years, and where do you think the woodlot
king of the block has gone without exception? To the jake, of course.

I've been thinking about throwing my hen decoy away.

CHAPTER 27

WHEN ALL ELSE FAILS

Tips for Working Stubborn Longbeards

- If you work a bird and he refuses to budge, do some legwork. Carefully work behind him and call from a different place. The change of direction will sometimes urge him to investigate.
- All calling options tried, get as close to a new set-up location as possible if you suspect he's with hens. Make it easy for him to get within gun range without too much travel on his part or yours. The drawback is that you may spook the birds. Use the ploy as a last resort and only in a 100 per cent safe area.
- Challenge the old, dominant hen with aggressive calling. This can work wonders – or drive the flock away. Long distance calls won't irritate her. That's why you'll have to move toward her and stake your claim. If she comes to investigate, so will the gobbler.
- If you've worked a stubborn bird for a day or two, chances are you've gotten a clue to his daily movement pattern. Get in the area early, set up and be patient. It's the way many old-time hunters, whose calling was limited to only a few putts, yelps or clucks every hour or so, approached the sport.

Jerry Peterson of Woods-Wise calls aggressively when challenging an older, aggressive hen.

- Do just the opposite. Call loud and long, hoping to fire up (and frustrate) a gobbler who is answering but not budging. If hen calls don't work, try imitating the gobble of a rival tom. Think of him as being a bull elk. The aggressive calling will eventually make him fight or run. Remember that on public lands the noisy gobbler (or you) may attract other hunters in the meantime. Be extra alert whenever you risk a gobbler call.

- Distant gobblers that continue to respond can often be approached by walking toward them and stopping regularly to call (Make sure you're wearing an orange hat when you do this.) In your calling, give the impression that a hen is coming to him. Get as close as the terrain will allow without being seen, then quarter away and retreat a shot distance. Call again, then sit and wait silently, at first, then continue calling. It's a toss-up as to what will work on any given day.
- Never use loud yelps before daylight. Move as close as you can to a roosted bird, then use subdued soft tree calls mixed with clucks and purrs. Don't call aggressively. Make him believe you're uninterested. Depending on a gobbler's mood, which may change from day to day, the curiosity factor may tip the scale your way.
- If you've fired up a bird but he refuses to move, abandon him for an hour or two. Seek other birds beyond his hearing distance, then come back and work him again. He may have completed his morning rendezvous by that time. If he hasn't been disturbed he'll probably be feeding and listening all the while.
- A sound that sometimes brings a gobbler into that magic 35-40 yard range is the scratching of leaves made by feeding birds combined with subdued feeding calls. Of course, movement alone may give you away when the chips are down so use the ploy cautiously. However, leaf-scratching is often easily done when seated against a tree.

Skip Stoudt of Emmaus gets ready to blow a barred owl call in a Potter County woodland.

- Gobblers in heavily hunted areas are the most difficult to bring in. Know the terrain and the listen for other hunters' calls. If you're interfering with another hunter, head in a new direction. Don't pursue any birds that you think are being heard by someone closer to it. Upon weighing all factors, either work the bird or abandon it in favor of safety.
- Don't write off a gobbler just because he's too distant; as far as a half-mile or more. If you're getting answers, even though they may be spaced by 10-30 minutes, you'll have gotten the message across to him that an available hen is about. Stay in one spot two or more hours if you must. Sometimes a stubborn bird will cover a lot of ground in a short time, even though earlier he'd refused to budge.
- The more you move about, the better the chance of bumping birds. On those welcome occasions when you see them before they see you, be rest assured you can't move. Dropping to your knees or a sitting position is often enough to declare your presence. Instead, freeze and hope the birds will move closer. Although it's occurred numerous times over the years, I recall only two

situations in which I've had shots at turkeys from a standing position. In one of those situations I stood like a statue for 15 minutes (My shouldered shotgun seemed to gain 40 pounds during the ordeal.) as a gobbler slowly moved to within 25 yards. It went home with me.

Bill Wary of Coopersburg demonstrates the "freeze position" used when turkeys see you before you see them.

CARL MOWRY
Butler, PA

Carl Mowry is president of the Pennsylvania Chapter of NWTF and serves on the River Valley Game Calls pro-staff. He's hunted turkeys for 30 years and "cutts 'n runs" in Butler, Warren, Forest and McKean counties.

CAN'T SIT STILL?
TRY THE OLD CUTT-'N-RUN

Depending on what type of area you're hunting, the "cutt-'n-run" style of finding turkeys might be just what you're looking for.

"Cutting," as you probably know, is the type of call most used in this technique. It is a somewhat aggressive series of clucks, usually offered loud and long. Of course, you're not limited to clucks as the technique also works with yelps.

I live in Butler County and the region in which I hunt holds mainly small woodlots of 1-4 acres surrounded by fields and houses. An area like this makes it harder to use the cutt-'n-run method, but it still can be done by moving around fields or developments.

However, 75 percent of my hunting is done in the vast Allegheny National Forest in Warren, McKean, and Forest counties. Cutt-'n-run works great in these wild areas. I usually use the technique if nothing is happening after the first couple hours of hunting.

The method is fairly simple and is based on covering lots of territory rather than setting up and waiting for a turkey to come to you.

I try to cover 150 to 200 yards on each "run," then stop and call. If nothing responds in 10 to 15 minutes I'll move on and try again. Depending on the area and how well you know it, you may choose to go around a ridge or over the top and down. There are no hard and fast rules.

It's important to move quietly and keep your eyes and ears alert for any turkey sounds. I can't say you won't bump turkeys off a ridge, but if you listen and are careful in how you move you can keep it to a minimum.

If you have not tried this before, give it a try. Instead of quitting after a couple hours of hunting due to lack of activity, cutt-'n-run. I'm convinced that it will also help you become a better hunter by understanding the movements of the birds and the terrain around you. It also gives you a chance to see the woods coming alive in the spring time and locating new spots to try another time out.

DALE BUTLER
Noxen, PA

Dale Butler is the owner of Red Neck Game Calls and president of the Red Rock Chapter of the P-NWTF. He does most of his turkey locating and hunting in Wyoming and Susquehanna counties.

LOCATING GOBBLERS AT SUNSET

It's early spring and I'm hoping to locate some gobbling birds. With the mornings short and working every day, I find roosting birds in the evening very useful. I like to locate as many gobblers as possible before the spring season begins and later use this knowledge for both scouting and hunting.

As soon as I get home from work I jump into the truck and head down the back roads along the mountains to some proven gobbling areas, discovered in previous years. At some sites I get out of the truck and often hear gobblers across the ravines and hollows without even walking.

Other times I will hike up the side of a mountain to get to the highest point where there's little interference from man and machines. Now I am ready to listen for the greatest sound of all -- the thunder of a grand, old gobbler.

Sometimes I will just listen in order to hear a gobbler down the ridge sound off on his own. Other times it is necessary to shock a bird into gobbling. Then I'll use different, loud sounds to stimulate the far-off birds.

My favorite locator is the sound of a barred owl with the cadence, "Who-cooks-for you, who-cooks-for-yawl." Many times I can hear the owl call echo through the hollows, sparking a tom to let loose with a reluctant gobble. I usually wait until the sun has gone down and the smaller birds are signing off for the night before using the owl hooter.

If I cannot get a response from the owl hooter I will use other locator calls such as a coyote call or a crow call. As a last resort I will use a turkey call, making clucks and yelps or a gobble. They'll all do the trick from time to time.

After hearing a gobbler, I like to pinpoint him as close as possible to his roost site. I prefer to know exactly which tree he is in, although this is not always possible. With that knowledge I can slip in on the bird before daylight and set up within 100-150 yards without spooking him.

JODY HUGILL
State College, PA
Jody Hugill is a field-staffer for Lohman Calls, Realtree camo and Fieldline. His favorite calls are the mouth diaphragm and slate. Jody carries a 12-gauge over-and-under and shoots 3-inch mag, No. 7-1/2 shot in the Centre and Huntingdon counties turkey woods.

SLOW DOWN, LISTEN UP FOR LATE SEASON HENS

Given one week in which to hunt spring gobblers it would be a toss-up for me between the third and last week. Many hunters have the impression that by late season gobblers have been figuratively "harassed and called to death."

Some have, but many simply shift gears and this requires some new tactics; like paying more attention to the hens. I live not far from Penn State and you'd be amazed at the number of students who hunt turkeys. They all seem to get it out of their systems – or have filled their tags – by the time weeks one and two have passed. That's when turkey hunting has always been best for me.

You need to understand that most hunters in this region only have gamelands or state forests on which to hunt and few people have honey-holes that others don't know about. By mid-season most hunters have given up, foliage is thicker and the patient hunter can move more freely and won't scare birds as readily.

That's when you start hearing what I refer to as the "call shy factor." Hens also go through cycles as they start to incubate and gobblers will now travel less as they learn where the scattered hens can be found when they leave the nest and head out to feed. In the last week or two of the season I will listen for hens and learn what they do in their own calling. I'll walk 20-30 yards and call and listen. Then repeat it.

Of course, you'll also stumble onto gobblers who will answer, perhaps only once or twice. When you set up on one, you'll simply need to be more patient than usual. He probably won't come running.

Be aware also that the increased greenery absorbs sounds. Calls from hens and gobblers won't travel nearly as far as they did on the first few days of the season. If you hear a hen or tom, chances are it won't be that far from you.

Late season hunting is a whole new experience. I've been so close to birds that I've have had to quietly work away from them before setting up. The whole Idea is to talk to the hens and get them to come your way – and hope that a silent gobbler is following.

LARRY SCARTOZZI, JR.
Greendell, NJ

Larry Scartozzi, Jr. has tagged a Pennsylvania
gobbler each year since 1988. He owns Kittatinny
Game Calls and makes a line of mouth calls.
He holds more than 120 calling titles across the
country's Northeast. Larry hunts in Susquehanna,
Pike, Monroe and Wayne counties.

MAKE THE RIGHT CALL FOR
ROOSTED TURKEYS

On my first day in the spring woods I watch closely if I've been lucky enough
to find a gobbler with hens. I pay very close attention to what they do and where
they head to after flying down. If I do not get to kill that gobbler the first day, the
information I obtain from the first hunt will increase my chances the next day.

In early May the turkey's sex drive is strong, so I usually get aggressive right off
the bat with cutting and yelping. I do not throw everything at them all at once. I
build into it. Whatever the turkeys respond to I'll keep doing. However, if a hen
yelps at me, I'll yelp back with the same call. If she cuts at me I'll cut back, imitating
everything she does. This can get a hen turkey very worked up and will increase your
chances of calling the flock in.

Often, changing your calls works. For some strange reason turkeys just like the
sound of a certain call. Which one it prefers on any given day is a toss-up. So try
different calls until you hit the right one.

A gobbler on the roost with hens in company will sometimes only gobble at his
hens and will not answer you. When they fly down, they shut up altogether. Wait
them out. Patience is very important.

Another tactic that works is breaking up the flock, like in the fall. I like to wait at
least a half hour before calling, unless they start calling first. This tactic works real
well when you have more than one gobbler in the flock. One of the gobblers will try
to beat the others back to the hens. When a tom has a lot of hens, I will leave him
and look for another gobbler. I will return to the first gobbler a couple of hours later
hoping some or all of his hens have left him.

I cannot emphasize the importance of good calling. If a poult can imprint to a
hen's voice while it is still in its shell, older turkeys can tell the difference between
a manmade call and the real hen. Hunters often do not put enough feeling and
realism in their calling.

For me, calling and getting a response is what spring turkey hunting is all about.

Longbeards in the company of others are usually more difficult to call in at season's end.

CHAPTER 28

IT AIN'T OVER 'TIL IT'S OVER

Don't Give Up Until the Clock Winds Down

The lingering minutes of the season were at once frustrating and exciting. Every 10 minutes or so for the previous hour my yelps and cackles had brought response from what I believed to be three gobblers holding their ground some 125 yards off in a brushy, Pocono Mountain thicket.

I'd first located the longbeards just before 11 a.m., an hour before the finale of the four-week season. Each of the trio had answered my yelps and clucks as I slowly wandered along a backwoods, gravel road closed to traffic. At about the same time I noticed posted signs between the birds and me.

I'd spent the morning on the border of a vast tract of Delaware State Forest land in Pike County, not far from Lake Minisink and Pecks Pond, and knew the wary toms had surely been serenaded by other hunters for the previous month and wouldn't fall easy prey to me.

Yet, it was crunch time and I felt lucky to have the chance for a last minute pinch-time tom. I set up within 50 yards of a deer camp's "No Trespassing" sign, hoping to lure at least one of the gobblers "across the border" before noon. Only once did all three birds bellow but the periodic gobbles from one of the crew provided hope that he might become sufficiently curious to set out in search of the lovelorn hen I was trying to imitate.

I set a pair of decoys on the edge of the untraveled gravel road in hope that the wary toms would move closer and see the imitations. Not only would calling in and bagging a gobbler in the final minutes of the season's final day be unusual and satisfying, it would also make a good story.

Still a good feeling

Sorry to say, it didn't happen, even though at one point the birds were only about 60 yards off, having worked their ways near to the property line, then promptly departed. My watch showed the official finale to be but a couple minutes off the noon closure and by the time I stuffed the decoys in my vest, the season was officially at an end. As I headed back to the truck I knew I'd given it my best shot and felt the moment was cause for celebration. I had to share it with someone. So I headed for a Pecks Pond diner for a sandwich and Coke, content that I'd given it

The final week's thick vegetation muffles sounds and hinders sight but some hunters swear it's the best of the season, as does JoAnne Zidock of Greentown.

my best shot. I shared my tale with the hunter sitting next to me but soon realized by the constant lift of his brow that he considered my "last minute" story exaggerated, if not a total lie. He told me he hadn't heard a bird all morning.

My experience aside, the final week of the spring turkey season brings with it a need to make some adaptations. As of now longbeards and jakes have seen or heard humans – or have been shot at – for 24 days. They're not only more wary than they were a couple weeks before but the greening and growing understory will have cut down on the distances they can see and hear.

The same goes for hunters who are often fooled into believing that gobbling birds are farther off than they really are, resulting in "bumping" them while trying to sneak closer for a set-up.

Hunters traveling into the turkey woods in the final week or so of the hunt must remember to don an orange hat while traveling through the woods. The increased vegetation brings hunters in closer contact with one another and raises the potential for accidents.

One advantage to late season hunting is that by mid-May most hens have already produced full clutches and are incubating, making them largely but not entirely off limits to gobblers whose libido has not been exhausted.

Unlike the opening days of trout, bear and turkey seasons when fishing or hunting produces the most action, the spring gobbler hunt is a season-long affair. Over the years I've filled as many gobbler tags in the final two weeks of the season as I have in the first two weeks.

Not everyone agrees, of course, and many hunters who hold the opinion that the best times are over by mid-season and stay home, making things even better for camouflaged diehards who don't give up until the stopwatch winds down.

Yogi Berra wasn't a turkey hunter, but in coining the now-familiar phrase "It ain't over 'til it's over," he hit home in the way baseball players and turkey hunters must continue trying to score no matter how late the season.

RON FRETTS
Scottdale, PA

Ron Fretts is a well-known member of the NWTF board of directors. He's been pursuing turkeys for 51 years and does most of his calling in Cameron and Greene counties. Ron's biggest gobbler weighed in at 23-1/2 pounds.

AFTER THE SHOT SAVOR THE MOMENT

Your bird is on the ground and your heart is doing flip-flops. It has been an exciting hour and 30 minutes.

What comes next?

An accident? It could happen.

After the shot, enjoy the moment. You have worked hard to get to this point, so take time and drink in this wonderful event. If someone is with you, it's time for a big smile, a handshake, or a hug. But before doing anything else, check with a good look to see that the bird is down for good.

Now stop, and make sure your safety is pushed back on. In the excitement, I didn't heed my own advice now and again and have run to the bird with my safety off. I don't do it anymore. This may be the easiest way to trouble; you could trip and your gun could fire in any direction, injuring you or your partner.

Remember always to stop, talk, and take some time. Let your heart rate come down. Then, with safety on, walk to your bird and savor the moment. I personally take a moment here to thank the Lord Jesus Christ for all He has given me, and especially for this harvested bird.

Occasionally, I have seen hunters with cuts on their hands from grabbing a flopping gobbler too soon. We know a head-shot gobbler is going to flop, so be patient and let him flop or put your foot on his head and push down on him to prevent his wings from thrashing about and damage to his feathers.

If you want to mount your bird, try to protect his feathers as much as you can. Don't pick him up and just throw him around, but carry him gently. Then, lay him on his underside to straighten his feathers, fold his wings and do anything to keep him fit for mounting.

While you are hunting, you may have to move and adjust quickly to the situation at hand. But after the shot – and it's struck its mark – it's time to slow down and savor the moment.

RONNIE JOHNSON
Cascade, MD

Ronnie Johnson has been making his personal line of turkey calls for
20 years. He also guides each spring in Texas, Alabama and Pennsylvania.
Ronnie lists Potter, Franklin and Adams counties as his springtime favorites.

TRY THE "LAZY MAN'S WAY"
OF GOBBLER HUNTING

Many times, particularly later in the season, pressured turkeys react to having
heard everything from A to Z with every turkey or locator call made. The more
hunters call, the fewer gobblers call back.

What's to do?

My choice is to become more of a stand hunter. I like to find a place where I know there are turkeys from having seen them earlier, including scouting time, and finding signs each day I'm in the woods. The hardest thing for some hunters to convince themselves at such times is that just because there's no gobbling doesn't mean that the birds aren't there. A lot of hunters give up when they don't hear any turkeys gobble and move on to the next spot. When walking or running and gunning, a lot of hunters push turkeys out of the area by being seen or making noise.

Remember turkeys can see 10 times greater then people with 20/20 vision and can hear four times greater then a person with perfect hearing. The natural cover and how quiet the woods are depends a lot on whether you can use that style of relatively static hunting or not.

I have had great success in first finding a spot with a good, wide and "comfortable tree" where I can sit and stay motionless for long periods without moving. Having a vest with a seat and back cushion is a must. Taking time to remove any rocks that you will sit on will also pay off later when the stones seem to grow. The smallest pebble can feel like the largest bolder after a period of sitting. Taking a couple minutes to level out and get a butt-protecting seat underneath you is very important.

The most important thing about hunting turkeys this way is being very patient? I try to find a set-up site close to food and water. Later in the season, when the hens split off from the gobblers to lay eggs and eventually incubate, they will need both. Always keep this in front of your mind.

I do very little calling; perhaps starting out with soft yelps, then switching to clucking and purring while scratching in the leaves to imitate turkeys feeding. I do this every 15-20 minutes or so. I listen closely, paying attention to all turkey sounds I have heard. Often a gobbler will come in silent and cluck to find you. Learn their language.

Sometimes in this situation the gobbler will be with hens so he doesn't have to gobble. His gobbling may be confined to assembling his hens. In late morning, if you're in an area where you didn't hear anything, consider that a gobbler may have heard you. Gobblers will stick with the hens for a time but as the hens move off to lay eggs he finds himself all alone. It's my experience that he will come back to where he heard hens in the morning and try to assemble them again.

An old turkey hunter once told me that turkeys have better legs then we do; so let them come to us.

In my seminars I tell people to abide by the three P's: practice, patience and perseverance.

I also tell them to sit and call light. Sooner or later the timing will be just right. The lazy man's way to turkey hunt?

Call it what you wish but for me it's become a winning late season technique over the years.

SECTION VII
The Grand Fall Hunt:
Different Time, Different Tactics

Guide Barry Haydt (right) of Danielsville and a client take a break following the harvest of a fall longbeard in Carbon County.

CHAPTER 29

BUSTING FALL TURKEYS
It's a Whole New Call Game

It's a whole new call game in fall when the green woodlands yield to pastels and the bubbly voices of lovelorn gobblers are replaced by kee-kees, yelps, clucks and putts of wary hens and their young of the year.

Pennsylvania, one of 38 states permitting fall turkey hunting, held its first-ever spring gobbler season in 1968. But fall hunting is steeped in a much older tradition. In the early 1960s and before, turkeys could be found only in remote pockets of the big forest counties in the northern tier. That limited hunting to autumn only. Spring hunting was not yet on the scene. Today the birds live

Greg Caldwell of Port Matilda was one of 31,100 hunters who tagged turkeys in fall 2003; his with bow-and-arrow.

nearly everywhere in the Keystone State and are autumn-hunted in all Wildlife Management Units except two zones in the far southeast.

Attitude adjustment

Hunting fall turkeys demands an attitude adjustment. The concept of declaring your position and purposely spooking a flock clashes sharply with the spring technique of sitting tight and not blinking an eye.

In a recent fall my son Andy, on his initial autumn turkey hunt in Pennsylvania's

Pocono Mountains, held off for a sure shot as a half-dozen hens and a jakes fed past at 50 yards. It wasn't until his kee-kees fell on deaf ears five minutes later that he realized he'd blown his chance.

"Oh yeah, I guess I should have run after them," he muttered sheepishly when we gathered back at camp in mid morning. "I was still in my spring mode."

Busting the flock is a standard technique in the fall hunt, as most participants know. But even this seemingly simple task doesn't always produce the desired results. Turkeys are social creatures and being away from the crowd for more than a few minutes makes them nervous, especially young birds.

Busting the flock isn't always as easy as it sounds. If the birds all fly or race off in the same direction, you've done little to separate them and you might as well move on to locate another flock – or try to skirt the birds you've run off and try again.

If the spooked turkeys run off, it may be worth a second try. If they all fly in the same direction, however, chances are slim that you'll see them again, at least not without some luck. Of course, the gamble you take in heavily hunted areas is that the birds you scatter may fly or sprint to someone else. But just because your efforts have resulted in another's benefit doesn't mean you need to give up hope.

The perfect bust

In Susquehannock State Forest in Potter County several years ago I ran toward a flock of a dozen hens and jakes which scattered in nearly as many directions. It was the perfect "bust." I found a suitable tree against which to hide and was about to slip the call in my mouth when a shot rang out from about 150 yards to my left. A minute later another two shots were heard from the opposite direction.

I decided to sit tight and wait another half-hour or so to allow things to settle. It wasn't long before I heard a "lost" call, then another. Within five minutes the twice-dispersed birds began filtering back, talking to one another – and me. The shots hadn't deterred the turkeys from their instinctive need to regroup.

The most effective calls include a series of low and high-pitched whistles plus the standard kee-kee run or "lost call" of varying volume and intensity. The idea is to make the lone birds believe someone from the flock is beckoning.

Continued calling, for five minutes or more, then listening for responses for a minute or two is a recognized technique. Birds of the year are the most gullible and will often respond to the pleas, including the older hens who want to bring the group together. Forget about gobblers who are more likely to be alone or in bachelor groups, not hanging out with hens and their young.

Find the food

As in spring, knowing your hunting area beats entering the woods hoping to stumble onto a flock. The advantage of the autumn hunt is that turkeys travel on their stomachs, moving with the soft and hard mast. Yet, hunters who expect to find birds where they found them in spring – or even during the October bow

Beech nuts are a favored fall turkey food throughout the state.

season – may be in for some disappointment. That's what biologist Arnie Hayden told his audiences during his frequent slide seminars.

"I know of birds that moved as far as eight miles to find mast," he told hunters. "When they find it they'll stay there until it's gone, then move on."

The way to find a turkey's haunt is through its stomach. Beech nuts arguably rank as the most highly preferred late summer/fall food followed in no particular order by black cherry, red oak, white ash, small acorns and bluebeech. Find a productive beech woods that hasn't been picked over and the birds will be there sooner or later. Black cherry is a predictable fall attraction as mast failures seldom occur with this common species. Red oak acorns, however, will vary in numbers from year to year.

Bluebeech, a common plant on Pennsylvania creek bottoms, is another turkey-attractor. The small nuts, which resemble "grape nuts," will hold birds in the area.

In my Northampton County woods, overgrown with sharp-needled barberry, turkeys come to eat the numerous red berries available in early autumn. They also like dogwood berries and will hop into a tree to get to them. In agricultural areas, add harvested corn and soybean to the menu. Find the food and you'll find the birds. That's pretty much the key to hunting the fall forests and woodlots.

Prime times

When's the day's prime time for encountering these birds?

It's been no particular surprise that over the years I've found the first hour and the final hour or two beating the "in between" time for locating the nomadic

Rob Poorman, Shirley Grenoble and Doug Marquardt head into the Sullivan County woods for a fall turkey hunt.

flocks. If you've found a roost and left it undisturbed it will serve as a magnet for a.m. and p.m. activity. Knowing where the flocks travel from bed to breakfast is a good place to set up and wait for the birds to work into the area.

In a recent November, in the final hour of the season, I set up in the same place I'd scattered a flock that morning, pushing the birds to a couple friends who each scored on hens. With only 30 minutes of legal hunting time remaining and the sky threatening rain, I watched a troop of 11 hens parade single file along a creek bed some 60-70 yards below. They were kicking leaves and pecking about on their slow return to their roost site a couple hundred yards upstream.

Knowing it was my last chance for a Thanksgiving turkey, I propped my Thompson-Center single-shot Encore against a tree, broke from the white pine against which I was pressed and ran down the hill toward the lead hen. Surprisingly, thanks to the help of gravity (At 60-plus years old I welcome all the help I can get.) I managed to intercept the flock, which may not have heard my approach due to the rush of water in the brook. Six hens broke off in one direction – toward the roost area – and the others retreated. I retrieved my shotgun, moved about 50 yards in the direction of the birds which had backtracked, then set up and offered a series of pleading "lost" calls. Three minutes later the first hen showed. I "whistled," she answered and I shot.

The season's end was a memorable one with little time to spare.

Those independent fall gobblers

While records of the percentage of Pennsylvania's fall harvest by sex are not available, my guess is that only about 20 per cent of the fall kill, if that, consists of longbeards and jakes. Many are shot with rifles at long distances. But shotgunners who need to get close to their quarry soon learn that techniques used for hunting large flocks don't work on wise toms.

Denny Gulvas of Tyrone, one of the state's best known turkey callers and the producer of a series of classic videos on fall hunting, rates autumn longbeards as "the toughest of any game."

"In fall, gobblers have nothing on their minds but finding food and staying alive and that's it." said Gulvas. "It's nothing like spring where they have mating on their minds and are gullible to calling."

Longbeards may join roaming flocks of hens for short periods but from midsummer through winter, they prefer to travel in small bachelor groups or alone. Only twice have I scattered a flock with an old gobbler among the "non-beards" and they didn't return when I set up and began my yelps and kee-kees.

"I don't think an old gobbler really cares if he has company or not," Gulvas has learned in his countless hours afield. "Anyone who breaks up a group of gobblers probably won't call them back again."

So what's a hunter to do if he or she is intent on harvesting a November longbeard? First, slow down.

"I do one old gobbler yelp every 45 minutes to an hour," said Gulvas.

"Then I suggest getting comfortable because if you really want only a gobbler,

Chris Rau of Unionville (left) and Paul Haydt, Danielsville, doubled up on a pair of hens in Hickory Run State Park, Carbon County.

Pennsylvania's fall gobblers are among the most elusive of all game.

you may have to sit a long time. Old gobblers don't do much talking in the fall and neither should you."

If patience is the key to successful spring hunting, it's doubly important in autumn.

Of course, fall offers the opportunity to take either hens or toms, increasing everyone's chance to tie a tag to a turkey's leg and bring home the makings for a wild Thanksgiving feast.

CALCULATING THE KILL

Record Harvest of 97,194 Was Set in 2001

Over the years the participation and success rates of spring and fall hunters have changed dramatically. In 2003, the most recent year for which spring and fall calculations have been completed, 246,821 spring hunters tagged 42,876 gobblers. Autumn's 211,965 hunters shot only 31,100 hens and toms, a plunge of 21.7 percent from the 10-year harvest average of 34,135, according to PGC turkey biologist Mary Jo Casalena.

"We can attribute that to rather poor mast productivity and more difficulty finding small flocks because they move about more seeking food," Casalena explained.

Since 2000, the spring harvest has exceeded 40,000 toms with a record-setting 49,186 in 2001. Add that to the calculated fall kill of 2001 (48,008) for an all-time record of 97,194.

Thee 10-year average fall harvest of hens and toms (39,707) remains well above the 2003 fall take of only 31,100 birds which Casalena attributed to "a poor spring hatch and poor mast crops."

The 2004 spring hunt yielded 42,876 toms, the third highest harvest in history.

Records show that average harvest success runs about 17 percent for both spring and fall hunters. By Wildlife Management Units, 2003 fall hunter success rates were best in WMU 1-B (28 percent) and lowest in WMUs 3 and 7-B (11 percent). For spring gunners, a mere 11 percent scored in WMU 7-A but 30 percent brought home gobblers from WMU-9B.

One troubling aspect of gathering reliable harvest information is getting hunter cooperation in reporting their kills. Turkey hunters don't even approach the shameful 50 percent or so reporting rates of deer hunters.

Casalena said report cards are filed by only about 15 percent of spring hunters and 32 percent of fall hunters. The postage-free cards are given with each hunting license or necessary info can be jotted on a postcard and sent to the PGC. Cards received and information from an annual game-take phone survey are used to come up with the calculated figures.

Casalena said the Game Commission is looking into a more workable system of reporting turkey harvests, possibly a toll-free number for calling in the kill. In Illinois, for example, I used a cell phone carried in my vest to report my gobbler within minutes after it quit flopping.

It's surely a subject requiring immediate attention.

Fall flocks travel with the available food and may cover two miles or more in a day.

KEN HUNTER
Muncy, PA

Ken Hunter is not only an accomplished outdoor artist but he also co-hosts Pennsylvania Outdoor Life, a weekly show on WNEP-TV in the state's Northeast. He turkey hunts in Montour, Lycoming, Columbia and Northumberland.

Here's Your Sign: Finding Fall Turkeys

Some turkey hunters, especially those new to turkey hunting, feel their preseason scouting trips are a failure if birds are not sighted.

Nothing could be further from the truth.

Much of my scouting actually occurs while looking for the ideal spot to bushwhack a deer in the upcoming archery season. While it's true that I will often see or hear turkeys prior to and during a bow hunt, it's also true that turkey sign discovered during these outings has led to a successful conclusion once the turkey season opens.

During a recent archery-scouting trip, close scrutiny of the forest floor revealed a fresh dusting area complete with tracks, several shed feathers and even some droppings. Each evening that I bowhunted the area I made it a point to check the dusting location and, sure enough, it was being used repeatedly. On opening morning of the fall turkey season I quietly approached a dusting area and I could hear scratching and "turkey talk" just over the ridge. An hour later I tagged my bird only 40 yards from the dusting sight. I had never actually seen a turkey there until that day.

Of course, in order to find evidence that turkeys are occupying a given area one must first know what to look for, then spend the time to seek out that evidence. You can't romp through the woods ricocheting off trees and expect to find sign. Look for scratchings or places where turkeys have raked back the leaves in search of food. Determine the direction of travel by checking to see which way the leaves have been kicked back. Try to determine how fresh the scratching is. Dusting areas are typically located where leaves have been cleared and a slight depression is made in the soil.

A turkey track with a middle toe four inches or longer is that of an adult gobbler.

Dry, dusty roadways also attract birds to dust. Turkeys will flap their wings and create dust baths in an attempt to rid themselves of lice and insect pests. These areas are often revisited. Dropped feathers are not only proof that turkeys had been there but they may reveal the sex of the bird. Gobblers have black-tipped body feathers and hens have a light brown-tipped body feathers. Droppings also reveal the sex of the bird. Gobbler droppings are J-shaped and hens form a small pile or glob. Of course, you can't neglect the obvious tracks found in mud along streams, trails and road edges.

DOUG MARQUARDT
Hughesville, PA

Doug Marquardt serves on the pro staff of
River Valley Game Calls and guides for River
Valley Outfitters in Sullivan County. He's a
member of the Muncy Creek Chapter of P-
NWTF and hunts Lycoming and Sullivan
counties and southern New York.

BREAKING THE FLOCK:
WACKY BUT IT WORKS

Scattering turkeys in the fall has proven itself to be a seemingly unusual but
very successful tactic. You can easily put the odds in your favor to bag your
Thanksgiving dinner by employing it.

I try to target a flock with several adult hens and their broods. If the opportunity
presents itself to actually "target" such a flock, your chance of harvesting a bird
greatly increases. My preference is to break up the flock in the late evening, just
before they go to the roost. To ensure a good break, I follow the flock until the
time is right; usually close to the roost site. Then I'll rush in at full speed, hollering
and waving my arms. This sends the turkeys fleeing in all directions without
enough daylight left to reassemble.

I will then stay at the break site until dark to ensure the flock doesn't get back
together. Then I leave and return to the exact spot the next morning, well before
daylight. The turkeys are usually eager to talk and get back together as day breaks.

Another method used by some hunters to bust a flock is to shoot in the air.
Although this procedure will probably do the trick, I do not practice it nor do I
recommend it. I feel that this is an unsafe method, especially on public land, and
disturbs other hunters nearby.

With several fall turkeys under my belt and several more in my buddies' freezers,
I have seen some unbelievable things happen in the turkey woods. One story
always comes to mind.

I was guiding a 42-year old first-time turkey hunter. He had many theories
and thoughts on how turkeys should be hunted. After the first hour hunting
with me he threw out everything he knew about hunting this wily creature.
Without finding any turkeys at daybreak, my client and I headed across a big
oak flat in search of a flock feeding. It wasn't long before I spotted a group
of birds about 70 yards out. I took off like a flash, hollering and waving my

arms. The turkeys scattered in all directions.

My client stood there in disbelief.

As I returned, slightly out-of-breath, he screamed "What in x@&?5# are you doing?"

I gave him a little smirk and told him "We're in – they'll be back."

When I'd calmed him down a few minutes later I explained this wacky approach to fall turkey hunting. He just scratched his head and said he'd never heard of such a thing.

In the next 45 minutes, after bagging a beautiful hen, the gentleman understood this strange strategy and the thought behind it.

But every time I looked in his eyes I could see his disbelief that everything he thought he knew about turkey hunting just went out the door.

ROB POORMAN
Muncy, PA

Rob Poorman guides turkey hunters spring and fall for River Valley Outfitters. He's on the ASAT Camo pro-staff and hunts primarily in Sullivan and Lycoming counties. His best bird is a 22-pound Sullivan County tom with a 10-1/2 inch beard.

THE FALL SET-UP:
HOW TO CALL THEM BACK

Once a flock of fall turkeys has been successfully scattered, it is time to set up for an attempt to call them back. You'll usually want to set up at or near the bust site, which is normally the center of the flock, before scattering. It depends on how the turkeys separated from one another as they raced or flew off.

Many times the turkeys will scatter downhill. In this instance, you should move closer so that they do not gather below you when you call them back. Don't be afraid of scattering the turkeys a second time if you find they are getting back together too far from your location.

If all or most of the flock scatters in one direction, move in that direction from the bust site in order to be among the birds should they return.

What's the recommended calling strategy?

In many instances, this may be the first time a family or brood of young jennies and jakes have been separated from the adult hen. Calling can and sometimes does begin immediately. In most instances, however, you'll have time to move and settle into your set-up and allow the woods to calm.

You'll want to listen at this point for the calling by either the young birds, usually in the form of the standard "kee-kee" run, or the yelps and clucks of the immature birds. Also try to pick out the dominant hen who will be giving assembly yelps. If you hear other birds starting to call, don't hesitate to join in. You'll want to sound like lost birds or, more importantly, like the dominant hen. Upon hearing the sounds of the scattered flock, try to imitate them.

If you have your sights set on a gobbler, use deep, raspy yelps, one to three at a time.

Remember, this isn't springtime hunting and you don't want to hammer away just because the turkeys are sounding off to one another – and, hopefully, to you.

◆ ANOTHER VIEW ◆

GREG CALDWELL
Port Matilda, PA

Dr. Greg "Doc" Caldwell is co-owner of River Valley Game Calls. He's hunted turkeys with gun and bow for 42 years and serves on the P-NWTF board. He hunts seven northcentral counties, often with his grandson Seth Marrara.

STRATEGIES FOR
FALL'S MIXED FLOCKS

Fall turkey hunting offers numerous flocks of mixed birds including brood hens with their young jakes and jennies. Therefore, the kee-kee run is the most frequently used autumn call. These are gregarious birds and making the high-pitched kee-kee sound is the "let's get back together" signal.

One of the keys to successful fall turkey hunting is knowing the flock type and matching its sounds. If I'm hunting barren hens (which have not been bred), I use a jake kee-kee, which is raspier and deeper in tone than that of a jennie. In addition, depending upon the area being hunted, I may slip in a jake gobble.

When calling toms or longbeards in the fall, I stick exclusively to one or two gobbler clucks and, perhaps, a single throaty yelp.

If I have roosted a known flock type and know the direction they usually travel, I may not make a sound. Instead, I'll set up and await the birds fly-down, giving time for a bird or two to eventually come toward me. This tactic is especially important when you are dealing with toms. If the flock does not come close enough for a shot, then a little scratching in the leaves or a soft cluck ore two just may be the ticket to entice them into range.

If that doesn't work, I then switch to a strategy most often used on mixed flock types; running at the birds and attempting to scatter them in all directions. I will then set up at the break site and begin using the calls mentioned earlier. If I'm uncertain which direction they will travel after fly-down, I'll scatter them off the roost. This is the method I almost always use with a mixed flock of roosted birds.

There are times, especially with mature gobblers, where extreme patience becomes critical. I may have to sit still for hours while throwing out just an occasional cluck. Many times the tom will return to the break area slowly and quietly. A ground blind works well in this case.

Another method is to return late in the day to a frequently used roosting site, then setting up and waiting. I recall a hunt where I had patterned a group of five longbeards. Knowing where they frequently went to bed, I set up a ground blind close to their travel area and waited (similar to setting up a tree stand in a buck travel corridor). Finally, after four days, the five gobblers came within bow range. At 17 yards, I harvested a 20-pound bird with an 11-¼-inch beard and 1-¼-inch spurs.

When locating gobblers in the spring, I'll often use locator calls such as owl, coyote or crow, but during the fall season, I speak only turkey language. I use calls that match the age and sex of the turkey I'm pursuing.

In the fall, if I haven't had any early morning success, I'll go on a "walk-about," checking for turkey signs while looking and listening for turkeys. I'll seek fresh turkey scratchings, droppings, and food sources. As I slowly move through the woods, I also will occasionally stop, calling softly at first then increasing the volume hoping to elicit a turkey response. Once they're seen or heard, I plan my flock-scattering strategy. After the break up, I will sit down for the "call back". If it is late in the day, the call back may be postponed until early the next morning and I'll once again work the roosted birds with a little more edge than the previous day.

SECTION VIII

After the Hunt:
Savoring the Meals, Mounts and Memories

A hunter takes a break following a successful morning hunt

CHAPTER 30

DOUBLE YOUR PLEASURE
Extend Your Turkey Season With a Video Hunt

The advent of the home video camera years back brought a new way of recording everything from capturing vacations on tape to recording the kids' birthday parties for posterity.

It also yielded a way for hunters to capture those memorable deer and turkey hunts on film. From VHS to Hi-8 and now digital photography, the sport of videoing your hunt has not only created a vast outdoor videotape business but also enables a hunter to show – not just tell – what happened during the day's outing.

Effective camouflage dress and a camo camera are vital to keeping a bird from seeing a videographer

The home videocamera today provides a new dimension – story telling with video – at turkey camps around the country. Instead of merely telling the tale at day's end, some hunters can now show it.

I've been privileged to appear in and film a few videos over the years and have been thrilled to share hunts with some true professionals. But capturing the sounds and sights of days in the gobbler woods is no longer confined to the few people who do it for a living. With quality video cameras having become more reasonably priced and more accomplished hunters extending their seasons by shooting video after they've shot their gobbler, it's no longer unusual to see camera packing, camouflaged sportsmen afield throughout the seasons.

Hunters who take their video production as seriously as they do the hunt can create professional looking programs that friends and family will cherish in the months and years to come while extending their turkey hunting time in the spring and fall woods as a bonus.

On the next few pages, Pennsylvania videographers Denny Gulvas, Ron Tussel and Don Jacobs share a few field tips on making your hunts into visual lifetime memories.

DENNY GULVAS
DuBois, PA

Videographer Denny Gulvas has been hunting turkeys for 36 years and frequently shares his knowledge as a seminar speaker. He's a 4-time U.S. Open Calling Champion who hunts in Clearfield, Cameron, Elk, Potter and Jefferson counties.

PRESERVING THE MEMORIES ON VIDEO

The nearly inaudible set of contentment calls proved just the right prescription.

Now, with an obvious air of confidence, the once reluctant gobbler inched closer and closer towards the compelling imposter set strategically just to our right. "Locked on" and ready for the inevitable, I was momentarily startled at the sudden flash of a speeding arrow streaking towards its intended target.

Within seconds it was over and the majestic trophy lay sprawled on the dew-covered leaves.

My son Cory had just taken his first gobbler with a bow. It was one of those special memories that will last a lifetime. The added bonus was that every second of that incredible encounter was recorded through the lens of my videocamera.

Each year more and more hunters are discovering that they can preserve their special memories through videography. As an outdoor video producer for over 20 years, I'd like to share with you a few tips that I believe can help you become a better videographer and forever capture those special days afield in pursuit of wild turkeys.

• First and foremost, work to become as familiar as possible with all your equipment. Strive to make the camera a part of you. Become intimately familiar with all its controls, switches and settings. Hours spent during the off-season recording different subjects under varying backdrops and lighting situations will pay big dividends later when that big gobbler makes his breathtaking appearance.

• Preparation of equipment well before each hunt is paramount. Good cameras eat power, so it's imperative to have several batteries ready and fully charged.

• Videotapes are relatively cheap but they are the heart of any recording. Start every hunt with an empty and full tape. Keep a backup within easy reach in case a quick-change is needed. Minutes tick by very quickly when dealing with a reluctant gobbler. I can't tell you how many times over the years a "once in a lifetime" shot slipped away because a battery ran low or a tape ran out.

• There are few things that rival the intense and exciting sounds of the turkey woods. Capturing every note is essential. All of today's top video cameras carry fine onboard microphones, which are great for general purposes. In the woods their drawbacks are many. For one, they pick up even the slightest hand movement associated with working the controls along with the annoying "hum" of the fast operating drum motor. Choosing a camera with an available external microphone jack that accepts a quality "shotgun" style microphone will do wonders for accurately recording every natural and exciting sound.

• As in hunting with a gun or bow, a good set-up is essential. Look a prospective area over well. Consider how far the viewer will be able to see an approaching bird? Are there any obstructions or blind spots? Correct lighting is important and strong backlighting can kill a scene.

• Lastly, always strive to keep everything smooth and flowing while filming. Avoid fast and erratic camera movements by panning slowly. A good fluid-head tripod is a worthwhile investment. Zooming the lens both "in" and "out" provide great subject detail but remember to zoom sparingly and as slowly as possible to keep the scene pleasing to the eye.

These are just a few of the many things that go into good wildlife videography. With a little skill, work and preparation, producing an exciting and interesting video of your hunt is within easy reach of anyone. Believe me, its well worth the effort to be able to relive all those "special" times over and over again.

(Author's Note: Gulvas has four turkey videos available and is currently working on his fifth. For details, email him at gobble@adelphia.net).

RON TUSSEL
Hawley, PA

Ron Tussel is the producer and host of Exploring the Outdoors TV Show and a columnist for Pennsylvania Outdoor Times. He's also a videographer for the Realtree Outdoors TV and video series. Ron hunts in Pike and Wayne counties.

COMMUNICATION THE KEY TO VIDEO SUCCESS

Capturing the excitement of a turkey hunt on videotape has grown to a passion for me. But when you're filming it's no longer a one-on-one sport. Add a camera and videographer to the mix and getting a gobbler up close can be difficult.

Hiding the camera and the videographer is a key element to successfully taping the hunt. Camouflage tape such as that available from Hunter's Specialties is used to cover all the shiny surfaces of my gear. I use camera covers made of Leafy Wear fabric. Sneaky Leaf and fake vegetation by The Cover System helps break outlines and hide movement.

The videographer – me or the person videoing me – dresses in full camo and sits directly behind or along side the shooter. Communication between the videographer and hunter is the key to getting great video. My shooter always has a wireless microphone on and I have an earpiece connected to the camera. The shooter can narrate the action, including which bird he is on, to be certain we're both focusing on the "right" bird before the shot is fired.

We also use voice activated, wireless two-way communicators. In the absence of electronic communications, we have used a simple thumbs-up and thumbs-down for the videographer to let the hunter know all is well. When the shot is imminent and I'm doing the video, I feather my hands off the camera so I am barely touching it. That prevents the reflex jump at the shot and the resulting shaky video.

Also, to make an above average show you need to shoot a bunch of cut-a-ways or video inserts. The tiniest details, such as a leaf blowing gently in the breeze, raindrops dripping off the brim of the hunter's hat or a silhouetted owl on a limb all help capture the essence of the hunt. So, too, are close-ups of the hunter settling up, calling, the safety coming off or the shell ejecting; all necessary to help the viewer enjoy and learn from the eventual, edited show.

Done correctly and artistically, cut-a-way clips enhance the finished product so the viewer feels that he or she is actually there, taking part in the quest.

DON JACOBS
West Wyoming, PA

Don Jacobs is the producer and co-host of Pennsylvania Outdoor Life on WNEP-TV. He appears both in front of the camera and behind it and must master all aspects of TV production. He hunts in Luzerne, Sullivan, Bradford and Wyoming counties.

A PRO'S ADVICE ON MAKING TV QUALITY VIDEOS

I have been known to preach the three basic rules of turkey hunting: Don't move, don't move and, whatever you do, don't move.

Now try practicing these rules while you're looking through the viewfinder of a video camera and a turkey is on the move. The best advice I can give a novice cameraperson is to treat the camera like a gun. Position yourself as if you were going to take the shot when the turkey appears. You can always get cutaway video later.

An experienced turkey hunter will point the gun in the general direction of the answering bird and so should the cameraman. Both the hunter and camera operator must be fully covered in camouflage. Every effort should be made to conceal the camera itself.

Most cameras offer both manual and automatic focus. If you are familiar with your camera enough to use the manual focus, remember that even the slightest movement in an attempt to focus the camera could catch the attention of the gobbler. Be prepared to keep your hands on the camera at all times in order to adjust the focus with the least amount of movement.

Start your taping early and at a wide angle mode until you capture the incoming turkey on tape. Once the subject is in view and on camera, you can slowly zoom in. Cameramen should learn to shoot with both eyes open, a certain advantage in keeping track of things.

Remember, a tight telephoto shot is harder to keep steady than a wide shot. If you use the autofocus feature on your video camera it may prevent you from zooming in on your turkey. Autofocus is designed to focus in on the nearest object to your camera. Trees and leaves may prevent the camera from giving you a clear video of the bird.

Autofocus is a good feature to utilize in an open field hunt because of the lack of obstacles. In woodlands, the autofocus is constantly adjusting itself and images flit back and forth from clear to blur.

Remember not to lift your head off the camera if the turkey walks out of the shot. Any movement can prematurely end the hunt.

Carrying a compact camera in your turkey vest enables on-the-scene snapshots.

CHAPTER 31

PHOTO FINISH

Record Your Hunt Like a Pro

As a full-time newspaper outdoors editor for more than 25 years I've received hundreds of photos of proud hunters and anglers posing with deer, turkeys, trout, elk, moose, bluefish and a variety of other fish and game. Trouble was, a full 80 percent of them lacked the quality necessary for newspaper use.

The shots varied from out-of-focus and under- or over-exposed prints and emails to photos taken in backyards next to the garbage cans or, the classic set-up, the beds of pickup trucks. Add to that the number one sin: wasting print space by taking the photo too far away from the subject.

Photos last a lifetime and it's worth a few minutes to set them up as professionals do for publication in outdoor books and magazines.

No matter if you have an

Even photos such as this one of Bob Danenhower of Lehigh County, taken in his backyard, are acceptable if a background is carefully chosen.

inexpensive point-and-shoot camera or a costly Canon or Nikon, a bit of creativity and having the camera ready will go a long way toward returning with plenty of "paper trophies" that your hunting friends will enjoy seeing.

Here are some tips for turning those bland, horribly-posed photos into works of pride.

- Carry your camera in your turkey vest with the intent of taking "on the scene" shots. Waiting until you get back home usually means setting up in one's backyard. If the yard has a hedge or conifer tree to fill the background it beats taking the photo with a patio, wash lines, house, vehicles or garbage cans in the shot.

The standard over the-shoulder pose, here shown by former Turkey Call editor Gene Smith, should include proper lighting on the hunter's face, the turkey and its beard.

• Clean the bird before photographing it. Use grass or moss to wipe blood from the facial area. Arrange the bird's feathers to show the iridescence that can be captured best on bright, overcast days.

• Shoot a few standard shots of the hunter kneeling behind his or her trophy and spreading the fan and wings. If it's a longbeard, also stretch out its beard. On such shots a fill-in flash, typically a feature of most modern compact cameras, will add a glint to your subject's eyes, take away the shadow from the hat and lighten the dark spots.

• Get your knees dirty. Lazy photographers will take shots of the hunter and gobbler on the ground, then shoot the photos from above. Try getting on the same eye level of the subject for a more interesting composition.

• Have your subject pose in the hat (blaze orange or camo) and clothing he or she was wearing while hunting.

• The over-the-shoulder shot is another popular pose. Hold the bird by its legs and lift it over one shoulder while slinging the gun on the other. Spread the wings so they sort of "cup" the hunter's upper body. Turn the subject slightly left or right so the beard is silhouetted against the background. Take poses with your subject looking head on into the lens, then snap a few of him or her looking out of the picture.

• Make a couple scenic type "snaps" of the hunter in the vicinity of the kill. Forget the shot of the person holding a turkey in outstretched arms with a death grip on its neck.

• Try a shot in which the successful hunter sits on a rock or stump and drapes the bird over the knee. One hand will be needed to balance the gobbler and the other to fan the tail. Again, spread the wings before pressing the shutter.

• Fence posts, fallen logs, rock outcrops, stone walls, low tree limbs, old barns, a blooming bush and other objects in the woods and fields make interesting backdrops.

Betty Lou Fegely poses with her Rio Grande gobbler draped over a knee with its tail spread

- Preen the bird, then put it on the ground, spread its wings and lay some of the gear you used – gun, shells, hat, calls and hunting license – on or around it. Shoot close-ups from directly over the set-up.

Professional photographers don't have a patent on making colorful and interesting photos. With some pre-planning and a touch of creativity, you too can turn a turkey hunt into a valuable photo-finish.

Outdoor writer Shirley Grenoble of Altoona set up against a stone wall background for her picturesque "photo finish."

BOB DANENHOWER, SR.
Orefield, PA

Bob Danenhower has bagged 13 gobblers over the past 15 springs and mounted hundreds more. His work as a taxidermist is rated among the best in the state. Bob has taken most of his spring toms in Lehigh County.

PREPPING YOUR TROPHY FOR THE TAXIDERMIST

You finally did it.

You shot a gobbler.

If you're like most of us who hunt, you can't wait to show off your hard-earned prize. The first thing you do is throw it in your truck and drive around to show it off. Heck, that's almost as much fun as hunting it.

But having been a taxidermist for 24 years, I have seen the disappointment in a hunter's eyes when he brings in his trophy after subjecting it to dirt, heat and maybe rain on his show-off tour. Add to that spoiled meat and feathers falling out.

Now he wants a quality mount.

Strutting your trophy around is great fun but there are some important steps you should take if you plan to eat or mount your turkey.

Field care begins as soon as the bird drops and the echo of your blast drifts away. When the tom hits the deck you click the safety, then walk out to find he is flopping, kicking, and batting his wings, often to the point that feathers start breaking and falling out immediately. Sometimes this nervous reaction to a head shot can go on for a minute or two.

When this happens, I usually grab the head of the turkey and hold him up and out away from my body until it is over. But I'm always certain that his needle-pointed spurs cannot hook you. I had a customer who had a tendon torn from his right hand by a thrashing turkey's spur.

Cooling off

Bird feathers and wings are awesome insulators that hold in the bird's body heat. When feathers and wings are collapsed against the body – such as when it's inserted in a carrying bag – it will hold its heat for several hours. That's not desirable.

It is very important to cool his internal temperature as fast as possible. The fastest way is to hang it by one leg in a dry, shady location that has some air movement. Spread out his wings and fluff up his chest feathers. This will speed up the cooling process. Allow him to hang for at least one hour. Now he can be frozen or refrigerated, always being aware that the feathers do not break or bend to hard.

To gut or not to gut

If you are going to have your turkey mounted, the best procedure is to take him to a taxidermist as soon as possible and allow him or her to gut your bird. The innards can stay in for approximately three hours before tainting the meat. I personally like to gut my turkey as soon as possible.

With two fingers, part the feathers in a straight line from the breast bone to its anus. This is approximately

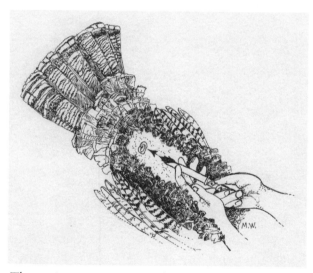

The gutting operation is a relatively simple process

4-5 inches. With a sharp knife make a shallow, straight incision between the anus and breast. Then reach in and pull out all the entrails, including the ones hidden in the upper body cavity. Lift the bird upright to drain any blood from the cavity. Keep paper towels handy to wipe off any blood that gets on the feathers.

Handle gently

If you freeze your turkey before going to the taxidermist, first assess the condition of the feathers. Your taxidermist needs to know if there are any broken wing feathers or missing chest feathers and if the tail is in good shape. This will help you and your taxidermist to determine the best pose. You surely don't want a mounted turkey flying to the left if he is missing a bunch of feathers from that side.

Getting a quality mount largely depends on how the gobbler was handled prior to taking him to a taxidermist

But that same turkey can still make a great mount; flying to the right with the bad side against the wall. A taxidermist cannot assess the quality of a bird when it is brought in frozen. It is up to you to make an analysis of its condition.

Freezing Your Turkey

Lay your bird on its back and smooth out its feathers, tuck the wings in tight to its body, wrap a paper towel around its head and bend the head and neck to one side of the body.

Now, slide the turkey into a large plastic garbage bag head first with the tail lying straight out. Push all the air out of bag and seal the end with tape. It is recommended that a second sealed bag is used.

Finally, place the protected carcass in a large freezer. Do not bend the tailfeathers. Your bird is safe this way for several weeks until you can get it to the taxidermist.

Be aware that most taxidermists have freeze-dried heads for replacement should your trophy's head have been too badly damaged.

A quality mount – on a limb, flying or strutting – will bring back memories for many years to come. Taking it to a talented taxidermist fit for mounting will help get the most from your investment.

Other displays

If you're not getting a full mount, you may want to consider a less-costly fan, spurs and beard display, a variety of which can be done by any taxidermist. Most taxidermists have photographs of previous works which provide ideas for a tail-spur-beard adornment.

A bit of advice: Be careful when removing the beard as cutting it off too short may eventually result in losing the hair-like feathers. It can also be pinched from the body skin with some care. To avoid rotting, treat the fatty end with salt or borax, preferably the latter. Keep the beard exposed to allow it sufficient time to dry, then clean it and add more salt or borax.

Also, be careful in cutting off the tail. Cut it short and you may trim off the fatty flesh holding the tail feathers in place. A simple operation of spreading the tail backside out and pinning to a board will dry nicely. Again, trim off excess tissue and use salt or borax to treat the fleshy parts.

Spurs also add class to mounts or can be used to make a necklace. Cut off the spur with a fine-toothed saw above and below the spurs' location. Then peel off the scales, remove the visible tendon and push the marrow out of the bone with anything from an ice pick to a piece of rigid wire. Then scrape the tail and soak it for a time in bleach or peroxide to make it white, if you so choose.

Or you can attempt the job yourself on some creations. Cabela's, Bass Pro, Redhead and other mail-order houses and larger outdoor sports stores have a selection of items for a variety of displays that anyone can make.

If you're not sure of your ability to get the finished product as you want it, remember your friendly taxidermist.

Cabela's Turkey Display Kit holds the spread tail and beard on a plaque

The lengths of both spurs are added and multiplied by 10.

CHAPTER 32

SCORE YOUR TROPHY

Do It Yourself With NWTF's System

The first measurement is the gobbler's weight.

"How many points," my friend Nick asked upon learning of my luck in bagging a big gobbler a couple springs back.

"Ten pointer," I glibly responded. "Had an 11 inch spread between the beak and the beard."

"I'm talking about your turkey, not your buck," he shot back with a smile.

Indeed, big bucks are ranked by points of all sorts, from Boone & Crockett and Pope & Young to Buckmasters and Safari Club International scoring programs.

But when it comes to measuring the "points" of a gobbler, there aren't any antlers to tape. The National Wild Turkey Federation takes over when a trophy has feathers and maintains an impressive list of more than 10,000 entries since its start in 1992.

The scoring system is based on weight, beard length, spur length and the total of all three items. State-by-state records are maintained for each. It's a relatively easy operation to score your bird although actually submitting an entry to the federation is slightly more involved.

Pennsylvania's best

For the record, the Best Overall Pennsylvania entry is a triple-bearded tom shot by Jeffrey Lange, Jr. in Bucks County in 2003. It scored 123.25 points, weighed 22.25 pounds, had an 11.25 inch beard and the longest spur taped 1.5 inches

Other notable Pennsylvania entries include Robert Wise's Greene County longbeard that pulled the scale to a whopping 27.75 pounds. Topping the list of longest-bearded birds is Michael Moore's Blair County tom with a 17.625 inch "paintbrush." The bird holding the biggest spur – two inches – was shot by William Haines in Washington County.

Of course, larger birds may have been taken in the Keystone State over the years but were not entered.

What's the Score?

Before you begin to score your turkey, be sure to note that all measurements are taken in 1/16-inch increments and converted into decimals. Even if you're not a mathematician you can easily figure things out by referring to the instructions and chart published here.

Be aware that a current NWTF member or another licensed hunter from the state must verify all measurements and attest to where the bird was harvested. Use a steel tape measure and an accurate scale.

To do your own measurements start by weighing your bird in pounds and ounces, then convert the numbers into decimals.

Next, measure each spur. Spurs must be taped along the outside center starting where the spur protrudes from the scaled leg skin to its tip. Add the two spur measurements and multiply the combined length by 10; the product is the number of points you receive for both spurs.

For the beard, measure the length and also convert it to decimal form. A beard must be measured from the center point of the protrusion of skin to the tip. Multiply the beard length figure by two. Convert to decimal and jot down the score.

If you have an atypical bird (with multiple beards), measure each beard and convert them to a decimal number, then add those figures and multiply by two. The product is your beard points. Add this to the weight and spur points.

The overall score of the three measurements is your grand total.

Birds with multiple beards are scored for total beard lengths, not just the longest.

Download a score sheet

Hunters must enter each turkey on an official entry application form, including certification that the turkey was taken by legal means, in the spirit of fair chase and not released for commercial hunting or confined within any artificial barrier.

Use the following list to convert each 1/16th of an inch to a decimal. For example, a 1-3/8th inch spur would be recorded as 1.3750. An 11-3/4 inch beard would be scored as 1.7500.

Conversions are as follows:

1/16 =.0625	1/8=.1250	3/16=.1875	1/4=.2500
5/16=.3125	3/8=.3750	7/16=.4375	1/2=.5000
9/16=.5625	5/8=.6250	11/16=.6875	3/4=.7500
13/16=.8125	7/8=.8750	15/16=.9373	

For additional information, download an official NWTF Wild Turkey Records Registration Form by logging onto www.nwtf.org.

All entries must be signed, witnessed and mailed to:
National Wild Turkey Federation
P.O. Box 530, Edgefield, SC 29824-0530

For Bagged Turkey recipe see page 243

For Turkey Salad recipe see page 244

CHAPTER 33

FARE GAME

Proper Field Care Leads to Prime Table Fare

By Betty Lou Fegely

You've hunted hard, the bird's in the bag and now it's time to think about the second reward of the hunt – a wild turkey dinner.

If you're far from camp or home and the late morning sun is heating things up, the first order of business is field-dressing. Along the with the entrails, also remove the crop, which stores food the bird's recently eaten before it passes into the digestive system. The crop contains fluids that will sour and give the breast a bad flavor if left in contact too long. The entrails will more rapidly taint the flesh, even during refrigeration, and removing them aids in the cooling process.

Troy Ruiz of Mossy Oak camo, who does much of his hunting under hot, Dixie skies, suggests hanging the bird by the head to drain out the blood as soon as possible after it is taken. When the entrails are removed, pack the cavity with ice so the bird cools from the inside out.

Many, but surely not all, hunters pluck the feathers, especially if the turkey is to be prepared whole. Otherwise, skin the bird or breast it with a fillet knife. After plucking or skinning, rinse the meat in cold, running water. There is no need to soak the bird in salt water, as your grandma may have done.

If the bird is to be frozen, double wrap it. Make the first layer a generous sheet of plastic or freezer paper wrap, to exclude as much air as possible. For the second layer, I prefer a resealable plastic bag on which I note the date and what's inside.

Marinating

Even though turkey meat has a delicate flavor of its own, I like to use marinades to enhance the taste. Marinating can be done in a zip-type plastic bag which makes refrigerator storage and cleanup simple. Marinades consist of anything from traditional salad dressings to liquids concocted from fruit juices, sauces, salsas or your favorite condiments.

Like fish, turkey meat will break down if marinated too long. My preference is to marinate for 2 to 4 hours. My favorite commercial marinades are Lawry's Teriyaki with pineapple juice and Mesquite with lime juice. Always marinate in the refrigerator or on ice. Warmth will aid in bacterial growth.

They're not butterballs

Turkey, like all wild game meat, lacks fat and care must be taken to not overcook it. If the turkey is young, pieces can be grilled, baked, broiled or deep fried. This is best accomplished by coating with a batter or marinating and coating with a crumb mixture. Grill or fry only a few minutes, depending on the thickness of the meat.

Overcooking will dry out even the best cared for turkey and is the main cause of failure in the kitchen. The longer you grill or bake a wild turkey, the tougher it gets.

For older birds, use the foolproof, self-basting, oven baking bag technique or bake it in a covered pan to keep the meat moist. Follow closely the cooking times on the baking bag box while keeping in mind that your wild game doesn't have the fat content of commercial butterballs.

From my kitchen to yours

There's always something special about a meal taken by your own hand. Doubling the pleasures of the bounty from a spring hunt is a treat in which the entire family can share.

Here are a few of the Fegely family's favorite recipes.

FRIED TURKEY BREAST

Take the skinned breast meat from one or more birds and cut against the grain into thin strips.

Marinade with the following:
1/4 cup Worcestershire sauce
1/2-3/4 cup Italian dressing
1 tablespoon hot sauce or Tabasco-type product
minced garlic - to taste
salt and pepper - pinch

Marinate turkey strips for 4 hours. Since turkey meat is very absorbent, do not marinate for longer periods of time.

Batter:
3 eggs
1/2 cup evaporated canned milk
1/2 tablespoon L.A. Hot Sauce or Tabasco like product
4 to 5 tablespoons yellow or Cajun mustard

Remove turkey from marinade and place in batter for about 45 minutes.

Into a plastic bag put 1 cup of flour and season to taste with garlic powder.

Take 6 to 8 turkey strips and place in plastic bag with seasoned flour. Shake to coat. Repeat the process, dipping back into the batter mixture and again into the flour.

Deep fry in hot grease 375-400 degrees no more than one minute.

This is great as an hors d'oeuvre. Serve with honey mustard dip.
Recipe contributed by Troy Ruiz, Mossy Oak.

BAGGED TURKEY

1 12-16-pound turkey - do not skin
Butter or margarine
Lawry's seasoned pepper and garlic salt
1 tablespoon flour
1 cup beer
1 apple, cored and quartered
1 orange, quartered

Rub turkey skin with butter or margarine, then sprinkle with seasoned pepper and garlic salt. Shake flour inside a large oven baking bag. Place bag into a baking pan and put turkey inside bag. Add beer, apple and orange. Close bag with nylon tie and cut 6 half-inch slits in top of bag. Bake at 350 degrees until done; 3-4 hours for whole bird; 2- 2 1/2 hours for breasts and thighs to be nicely browned. Serve with cranberry sauce, your favorite stuffing, and green beans or peas for added color.

TURKEY FAJITAS

1 pound boneless turkey breast
1 medium red onion, sliced
2 tablespoon olive oil
2 tablespoon lime juice
2 bell peppers cut into strips
garlic salt to taste
3/4 cup salsa
8 flour tortillas, warmed
1 cup shredded mild cheddar cheese

Combine turkey and first 6 ingredients in a large resealable plastic bag. Marinate 2 hours or overnight in a refrigerator. Grill turkey breasts about 7 minutes. Place sliced vegetables and marinade in a sauce pan and heat until vegetables are slightly cooked. Cut turkey breasts into thin strips and divide among warm tortillas. Drizzle with hot marinade. Sprinkle with cheese. Roll up and serve immediately. Serves four.

WILD TURKEY SOUP

1 turkey carcass, meat removed
4 medium carrots
1 small cabbage
3 stalks celery
1 teaspoon salt

1/2 teaspoon pepper
1/2 teaspoon poultry seasoning
1 (16 ounce) package macaroni, cooked
Leftover stuffing and gravy (optional)

Break carcass into pieces and place in an 8-quart soup pot. Cover with cold water (4-5 quarts). Bring slowly to the boiling point. Remove film. Cover and simmer 2 hours to loosen meat. Set aside. Chop carrots, cabbage and celery in a food processor. Add chopped vegetables, salt, pepper and poultry seasoning and any remaining stuffing and gravy to the broth. Simmer 2 hours. Adjust seasonings. Add macaroni.

Best served the next day. Makes 8-10 servings.

Recipe contributed by Rob Keck, NWTF.

TURKEY SALAD

Use drum sticks, wings and backs. Place in a pot with onions, celery, bell peppers and cover half with water. Boil until meat is tender and falls from bones. Strain and remove meat.

Chop by hand or use a food processor:
2 green onions
2 bell peppers
2 tablespoons minced garlic
3 shallots
salt and pepper to taste

Add chopped meat to vegetables. Mix 1/2-3/4 cup mayonnaise to the mixture and serve with tomatoes, lettuce and crackers.

Kelli and Mike Kostick of Cherryville toast a meal provided by nature's bounty.

CHAPTER 34

MOREL MAGIC

More to Hunt Than Just Gobblers

Of course, spring is also the time woodland wildflowers burst into bloom with everything from trilliums to dwarf dogwoods and mayapples coloring the woodland floor. For a growing number of turkey hunters, however, the grand spring sport offers an additional challenge – hunting morels.

On a May hunt with several friends a couple years back the gobblers remained close-mouthed one misty morning but an unlikely treasure saved the day. It was prime time for mushrooms and nearly everyone returned for lunch with a gamebag full of them, making our evening feast ultra-special. The camp

The cherished morel mushroom is a bonus find on any trip.

cook prepared them in several ways, including stuffing them with cheese and, the tastiest recipe, breading them then frying in butter. Cholesterol aside, they were fit for the proverbial king and every bit as enjoyable eating as they were in finding.

Both texture and taste of the morel rank it as one of the best (many mushroom hunters say it's THE best) of the edible fungi. Furthermore, this springtime find is easy to identify. Its appearance has been described as resembling a pine cone, although ridges and pits, not scales, compose its surface. Its flesh-colored head is attached directly to a stalk, which usually grows shallow in the soil.

A day or two following rain and sunshine the readily identified fungi typically show their 4-8 inch, golden heads. A friend, who remains close-mouthed about his annual morel hunts, usually done as he seeks gobblers, advises locating old apple orchards, moist oak ridges and burns. I've never pressed him for more specifics on their locations.

When readying morels for the table or for freezing, keep in mind that the head of the morel is hollow, providing a ready site for insects to hide. Each head should be sliced open and the inside thoroughly rinsed. Whether breaded and butter-fried or sliced and served atop a New York strip steak, the morel will be cherished by all.

Morel pickers who stumble upon mother lodes of the gourmet fungi are known to keep tight-lipped about their locations.

"You won't find someone coming up to you and saying: "Hey, I found a bunch of morels. Why don't I show you where they are so you can tell all your friends?" a turkey hunting companion of mine chuckles. "Only I know where my honey-hole is located."

Small openings in woods such as mini-plots and ATV trails will attract turkeys for greenery, seeds and insects.

CHAPTER 35

PLOTTING FOR WILD TURKEYS

Food Plots Aren't Just for Deer Anymore

If you're a property owner or belong to a hunt club, "calling" turkeys should begin long before the first gobble of spring. Although deer have stimulated most of the interest in food plots in recent years, poults and adult hens and toms also benefit from the specialized plantings.

Improving habitat to lure turkeys to your holdings – and lure them there time and again – demands setting the table with nutritious fare. Turkeys are as much field birds as creatures of the woodland and they spend considerable time in clearings.

"Wooded tracts lacking openings hold less appeal for turkeys," says Bobby Maddrey, director of Land Management Programs for the National Wild Turkey Federation. "My recommendation is to create a bunch of small plots rather than just one or two large openings. In larger clearings a big portion of the middle often won't be utilized."

So where do we start?

"Some property owners pick their worst land to create openings," said Maddrey. "It must be considered that good soil is the key to a good food plot."

Generally, moist areas are more desirable than dry hilltops. Low sites with poor drainage may be equally unfavorable, creating wet, muddy conditions. Determining the top sites for creating clearings is best performed with aerial photos, particularly infrared images. The bird's-eye views take the guesswork out of mapping desired openings in relation to timber stands, ponds, streams, fencelines, roads and natural openings.

Maddrey recommends creating clearings no smaller than a quarter acre so that nearby trees can't drain the nutrients and shade the plantings.

Of course, quality seeds and seedlings are prerequisites to successful food plots. Choosing wisely and paying a bit more at the start may mean fewer failed plantings and saves both money and time. Biologist Maddrey recommends "some form of clover and legume grass mixture" as a basic, turkey-attractor. Deer will eat the faster-growing grass and leave the slower-growing clover.

Suggested grasses include rye, wheat, timothy, bluegrass and orchard grass or mixtures of each. Not only will turkeys feast on the grass but they'll also strip the

Clover is a deer and turkey favorite.

seed heads as the plants mature. Several clover choices are also available with the red clovers – such as crimson clover with its striking, red flowers – the most desirable. A bonus is that insects are attracted to the flowers, in turn providing important protein sources for poults whose diets are composed of 90 percent insect life.

Across much of the South, chufa is the best-known turkey food. The grassy plant produces nut-like tubers which the birds scratch from just beneath the surface. It will grow northward to Pennsylvania although it has yet to gain popularity. It's an excellent winter food in regions with 100-110 day growing seasons. Whenever and wherever the ground freezes, however, turkeys may find it impossible to scratch up the chufa when they need it the most.

Finally, liming and fertilization of the plots is a must. County extension services will test the soil for a few dollars and offer advice as to the amounts of each necessary.

What about winter foods?

"Consider the year-around needs of the turkeys on your property, not just the warm-weather needs," Maddrey stresses.

Throughout any wild turkey range, but especially in the North, mast-producing species are important survival foods. Crabapples, white ash, hackberry, staghorn sumac, barberry, raspberries, persimmon, sawtooth oak and others provide acorns, seeds, fruits, berries or buds. They're especially important fall through winter.

Maddrey also tells landowners that not all openings should be planted.

"It's a misconception that you must plant something in every clearing," he said. "There's a ton of different stuff that will come up when the sun is able to pour in."

As an example, he points to dandelion – an undesirable weed in most situations – as being a good turkey food. Other weeds, herbs, grasses and forbs will also thrive following bush-hogging or disking a small section of open woods.

In such openings, mast-producing trees and shrubs should be left standing but it may be necessary to thin them so that sunshine can reach the woodland floor. Cutting competing vegetation around beeches, oaks and other mast-producers is also wise, says Maddrey.

If turkeys find it necessary to work their ways through dense vegetation to get to a clearing they may not come at all, says Maddrey. Their natural fear of predators and the inability to see very far may make them travel-shy. Bush-hogging pathways to the clearing sights may help but predators will also frequent these ambush routes. Also consider widening existing gravel roads to provide edge-effect habitat. Disking, then scattering grasses in such sites, makes for great brood habitat.

Finally, don't seek a quick fix, advises Maddrey. "Make an overall plan and determine your goals based on the money and time you have to spend."

Even before the first seed or tree is planted, an analysis of a tract's overall habitat requirements should be performed. Determining which parts of the puzzle are missing is the first step in any successful turkey food plot project.

For specifics on wild turkey plantings, contact the National Wild Turkey Federation's Project HELP (Habitat Enhancement Land Program) by calling 800-THE-NWTF.

Poults and adult birds frequent grassy clearings throughout the year.

STEVE TRUPE
Columbia, PA

Steve Trupe is a consulting wildlife biologist and habitat management consultant. He also runs a deer camp in Potter County. Steve pursues turkeys in Huntingdon, Bedford and York counties.

FOOD PLOTS FOR DEER ALSO FAVORED BY TURKEYS

I plant close to 50 acres per year in wildlife food plots for myself and my clients. With the exception of an occasional chufa or sunflower patch, I seldom plant anything specifically for turkeys.

It just so happens that turkeys are quite adaptable and not very finicky when it comes to diet. They will be attracted to and eat just about anything you plant for deer or upland game birds. Think of a wild turkey as a cross between a white-tailed deer and a ringneck pheasant and you can't go wrong in your planting choices such as alfalfa, clovers, vetches, peas, beans, grasses and all the cereal grains (wheat, rye, oats, triticale, corn, sorghum) as well as the brassicas (turnips, rape and beets).

In most cases they will eat the foliage, seeds and the insects that the plantings attract. For winter foods in areas of deep snow, standing corn and sorghum are great. So, too, are fruit producing trees and shrubs that hold their fruit well into the cold season including hawthorne, crabapple, cranberry, bittersweet and barberry.

In areas where the habitat is a good mix of large mature woodlots, crops and pasture lands, turkeys need little help to thrive. It's in the big woods regions of Pennsylvania that food plots can be very beneficial. Turkeys need open grassy areas for their young to forage for insects; usually few and far between in forest situations.

Plant any openings you have – logging roads and wide roadsides, log landings, powerline and pipeline rights of way – in perennial legumes such as clover. The best planting time here is late winter or early spring when you can "frost seed." That means broadcasting the seed on frozen ground and let the freezing and thawing action work the seeds into the soil at a time when moisture is abundant.

Plot size doesn't matter as much to turkeys as it does to deer. Gobblers, hens and poults will use large fields as well as small log landings when and where there is adequate visibility around small clearings. Thick brush allows predators to get too close and may make the birds wary.

Plant the aforementioned fruit producers near spring seeps that stay open in winter or establish them low on south facing slopes where the snow will melt first and turkeys have access when they need it most.

In early spring, plant trees and shrubs for their best chance of survival. My suggestion is to take a detailed look at your property and the surrounding habitat to record what components are lacking for brooding areas and food sources. Then, spend the bigger portion of your budget and energy improving that aspect of your property.

When in doubt, consult a professional to visit your deer and turkey hunting grounds and make recommendations.

(Contact Trupe at 717-684-2230.)

SECTION IX

Passing It On:
Bringing Others Into the Outdoors

Ray Smith of River Valley Outfitters called this 2004 spring gobbler for Joseph Leppert on the youngster's first time out.

CHAPTER 36

A KID'S FIRST TIME OUT

Make a Test Run Before That All-Important Day

Sometimes it pays to look back, despite Satchel Page's sage advice to the contrary.

I vividly recall my first morning in the whitetail woods; my dad an acorn's toss away beneath a towering oak. It was an exciting time following a sleepless night in which I'd mentally rehearsed exactly what I'd do when a hat-racked buck strolled along the trail 60 yards down the ridge.

It never happened. Save for several does that sprinted by in mid-morning, raising my blood pressure, the action that inaugural deer hunting morning was minimal.

But the memory remains crystal clear. The same holds true for the first gobbler I ever saw fanned and strutting. I never got a shot at him but I did learn not to move when he's looking.

Such are but a couple cherished memories, among others, of my country upbringing.

The author's stepdaughter, Kelli Hoats – now Kelli Kostick – shot a Lycoming County gobbler her first time out at age 12, back in 1984.

Taking kids – indeed, any friend or relative – hunting for the first time is a colossal responsibility, no matter if you're after whitetails or fantails.

With years of experience under my belt I served as mentor to my two sons, several neighborhood kids and, later, a 12-year-old step-daughter. My approach was to recall my first days – first years, actually – in the field along with the naiveté and the occasional fears and unknowns with which a youthful mind approaches any new challenge.

Field trip tips

If there will be a young hunter at your side this year, consider these

John Annoni of Allentown directs Dianna Martin of Camp Compass on safe handling and shooting a shotgun.

suggestions leading up to the first time out in the turkey woods.

• First off, Pennsylvania requires the completion of a hunter education course before a license can be purchased. After the course, review the lessons and reinforce the safety aspects covered with the young hunter. Stress the responsibilities inherent in carrying a firearm and the importance of a clean, humane harvest.

• Spend time on the range throughout the year familiarizing the newcomer with his or her firearm's operation and its safe handling. A shotgun can be

Taking a child under 12 (sans gun) on an actual hunt is both exciting and educational, depending on the maturity of the child.

Andy Sawka of Northampton County provides close-up attention to his son Michael on a fall turkey hunt by sitting the young hunter between his legs.

intimidating with a natural fear inherent in firing it the first few times. Flinching is a common affliction. Problem is, the shooter often doesn't realize he's flinching and therefore shooting inaccurately.

Try this: While shooting at a turkey target to teach the lessons of patterning, sneak a spent spell into the chamber if you believe your student is flinching, then watch his/her reaction when the trigger is squeezed but no shot rings out. The lesson is quite revealing – to student and mentor alike.

• Take your new partner to the hunting grounds where you'll be spending the opening day of the season. Preferably, make it a mock hunt by getting out there at daybreak in hopes of finding roosted birds. Use a locator call a few times but forget about (even though it's very enticing) using turkey sounds. Gobblers and hens will become educated soon enough when the season gets under way. If it is feasible, taking a child on an actual hunt should also be considered.

• In Pennsylvania it's mandated that a licensed adult be aside the young hunter and "close enough that verbal guidance can be easily understood." That's no problem as "team" turkey hunters are typically side-by-side anyway. Indicate "out of range" shots and unsafe shooting methods.

• Do a set-up as if a turkey has been heard. With first-time-out hunters, I prefer to set up with my back against a large tree and my student sitting between my legs. That enables whispers to be heard and the mentor has control. It also aids in making necessary hand or gun movement. This also a good time to work

It's never too early to buy a kid a box call and show him or her how to use it.

on estimating distances, a judgment call with which even adults have problems. Carry a turkey target or better a decoy with you and set it up at different locations from your set-up site. Have your student guess at the yardage, then measure (or pace) it off.

• Also act out the recommended set-up of the mentor sitting with his or her back to a tree and the boy or girl sitting between his legs. This enables easy communication and reduces unnecessary movements. If a turkey comes in, the "guide" sees exactly what his partner sees.

• While in the woods, discuss the importance of sitting still, speaking in hushed tones, walking in leaves or snapping twigs, all of which can spook a wary gobbler.

• Include your youngster when patterning his or her shotgun. Use a sandbag (preferably filled with beans, not sand) or a shooting vise and provide instruction on how the pellets leave the muzzle and form a cone-like pattern which gets larger the farther they get from the muzzle. Take shots at 10, 20 and 30 yards and compare results. Also include offhand shots and shots at targets while in a seated position, emphasizing the proper way to prop a gun on a knee as a gobbler moves close.

Make the entire pre-hunt experience a positive one; not threatening or too complex. It will assure confidence, anticipation and familiarity on that memorable first time out.

And it just might make the difference between your student becoming a one-time-hunter or a "lifer" ….. like us.

Get involved in a youth turkey hunt

One of the best ways to "pass it on" is to find a partner for the annual Youth Turkey Hunt held the Saturday prior to the opening of the spring gobbler season.

The event for kids 12 to 16 is coordinated by the PGC but the hard work is done by sportsmen's clubs across the state.

One call and guide service company – River Valley Outfitters in Hughesville – sponsored an essay contest with the winner getting a free hunt and an ASAT Vanish Pro 3-D camo package and turkey calls from River Valley in spring 2004.

"We were pleased to be a sponsor," said River Valley owner Ray Smith. "There's nothing more satisfying than taking a beginning hunter into the woods with the guidance that leads to a successful turkey hunt."

For more information on the contest and River Valley's services, log on to www.rivervalleygamecalls.com. For general info on the statewide youth hunt access the PGC's website at www.pgc.state.pa.us.

Doug Marquardt, Ray Smith and Rob Poorman of River Valley Outfitters (back row) and their young hunters Danny Dochinez, Joseph Leppert and Zack Nessa pose following a successful Sullivan County hunt. Photo courtesy of the Pennsylvania Federation of Sportsmen's Clubs.

JOHN PLOWMAN
Mechanicsburg, PA

John Plowman is a retired legislative liaison for the Pennsylvania Game Commission. His best hunting memory is a recent one thanks to his grandson's first tom. John hunts in Sullivan, Dauphin, Wayne, Lycoming and Greene counties.

A TOM FOR JAKE

It's 6:15 a.m., April 24, 2004 on Blue Mountain.

Jacob "Jake" Gettys and I squint through our facemasks scanning the maze of possible approach paths for any movement. Ten minutes ago one bird closed in but slipped by unseen and kept going. But his buddy was still out there and closing in, tracked by his thunderous gobbling.

The occasion marked the state's first spring gobbler youth hunt, a time for my own 50 years of acquired hunting experience to begin transfer over to my 12-year old grandson Jacob, aptly nicknamed Jake. This morning he's set up in my lap, a new 20-gauge Remington 870 on safe and steadied in his clutch. For Jacob, this was the culmination of practicing woodsmanship, shotgun patterning, sportsmen's ethics, respect for wildlife and developing a hunter's fundamentals.

Ten minutes earlier two good toms were gobbling above us, yet we were too close and pinned down at the edge of the newly logged-off bench that would become center stage. Thankfully Jake's pretty savvy about the outdoors, even when things don't quite turn out as desired.

We heard him drumming first, out of view beyond the clearcut's perimeter, yet he was working his way through the snags and treetops toward us. The huge tom popped out in full strut for the decoy at 40 yards, then angled closer. Jake reacted calmly to my whispered instructions, sensing the bird's unforgiving eyesight peering through us like x-ray vision. Now all this 20-plus pound bird had to do is slip out around the one treetop between us. Then he'd wear Jacob's tag.

Soft purrs with the Rohm Brothers diaphragm triggered the necessary neck stretch. I whisper to take the shot if he thinks he could put the gobbler down. It was an ideal opportunity at 15 yards, yet Jacob calmly declined his only shot, unconvinced of a clean kill with his 20-gauge. Mr. Longbeard soon "made us," reversed course and "putted" back the way he had come, offering a memorable show but not the shot.

Another "opening day"

Fast-forward to 5:30 a.m., May 1. It's the Pennsylvania gobbler opener. This time we're on a favorite ridge in Sullivan County's scenic Endless Mountains, replete with a flock of pre-scouted gobblers. Off to a slow start with nary a bird calling on roost. After our frustration a week earlier we didn't need a silent spring, too.

Jake and his dad Jeff set up just below the summit of Madi's Notch. Despite past successes here with these patterned gobblers, by 7 a.m. it was obvious nobody was in a very talkative mood. The perfect if uneventful morning was slipping away, with enthusiasm fading as visions of blueberry pancakes started looking better than chasing nonexistent turkeys in the 80-degree heat.

But one trick remained; a distant place the turkeys could have slipped into. Jeff and Jake took off for our last-chance location, knowing to float hunt along the top, call easy and wait for developments at each stop until reaching our vantage point overlooking the remote hollow.

Once in position with Jake poised for action, Jeff eased over to the edge and ran his longbeard slate for effect, immediately triggering gobbles from multiple toms below. Hope at last.

This time the birds were receptive and far downhill in the grapevine and hemlock coverts. Our intimate knowledge of the property plus plain good luck and guesswork had worked its magic.

With Jake set up to allow the maximum swing and a safe shooting zone, Jeff crouched behind him and fired up the birds to bring them out of the hemlocks. The rest of the story is anticlimactic. First, to dash uphill was bachelor number one; a 17-pound trophy with a five-inch beard was smoothly dispatched by Jacob via a perfect headshot, again at 15 yards.

Like he's done this for years.

That resounding shot and the hunting tradition it initiated will echo in those hills forever. It will be a lasting memory for Jake. Our new hunter officially tagged out opening day – a fitting tribute for the 12-year wait and the best of hunts for all of us.

◆ ANOTHER VIEW ◆

John Annoni (far right, bottom) with the Camp Compass group.

JOHN ANNONI
Allentown, PA

John Annoni is a school teacher whose out-of-school efforts have made Camp Compass recognized across the country. He carries his scoped Remington 870 in the woodlands of Lehigh and Susquehanna counties.

CITY KIDS TAKE TO TURKEY COUNTRY

When you think about the traditional introduction hunting programs for today's kids, a program I began in 1996 is considered unique.

I say this is because my approach to kids and hunting is probably as odd as a double-bearded tom. It's called Camp Compass Academy, a non-profit organization based in Lehigh County. It continues as my vision to provide children the opportunity to experience hunting, fishing and the all-outdoors.

Our main focus is on city kids who normally would have no other exposure to hunting and fishing. It's one thing to have families introducing their children to the outdoors, but to teach inner city children to pursue critters is a totally different game. It's not been without its rewards. The academy has been recognized locally, nationally and internationally. I've also appeared on an array of television and radio shows and often speak at outdoor related functions.

But I couldn't do it alone. Through my volunteer staff we have introduced hundreds of children to the outdoors by using our effective, multi-level approach. Boys and girls of different ages, races and religions have experienced first hand the powers of outdoor engagement.

In 2004, thanks to the new Pennsylvania Youth Turkey Hunt, I was able to enjoy my first opening day of turkey season in a non-mentoring scenario. It was the first time in eight years I was able to hunt with friends. For the past eight years I've given my opening day to kids.

This year's youth hunt found us at Full Fan Lodge in Susquehanna County with owner Brian Post and his array of turkey guides, who offered to host eight Camp Compass members. Two toms were harvested, one was missed and everyone saw or heard gobblers.

We all considered that it would have been a winning day for adults, yet alone city kids. Each of our hunting teams consisted of a turkey guide, child and trained hunt-mentor. Our successes in the field and outside of it speak volumes about the attitude with which we hunt and the kind of dedicated people Camp Compass Academy attracts.

We call it family – an outdoors family.

For more information on Camp Compass, contact John Annoni at info@campcompass.org.

KARL POWER
Murrysville, PA

Karl Power is the outdoors editor of the Valley News Dispatch and an outdoors radio broadcaster. He's been hunting turkeys for 20 years, mostly with his two daughters in Westmoreland, Indiana and Potter counties.

Jessica, Karl and Maggie Power

DAUGHTERS' LONG-AGO LESSONS STILL PAYING OFF

Having raised my two daughters single-handed from the time they were in diapers, they were destined to become hunters and anglers. Both of them – Jessica and Maggie (now in their early 20s) – quickly became successful deer hunters and continue testing their skills with each taking two deer (buck and doe) each season.

Convincing them to begin turkey hunting took some coaxing, but they agreed to give it a shot. The spring season was the best way to go as that early-morning gobbling will trigger excitement in anyone. However, both have also enjoyed the fall season – having taken both hens and fall gobblers.

Sitting completely still was the initial issue for both daughters. Neither one understood how much more visually alert turkeys are as compared to deer. I found that the best way to keep them still was to equip them both with comfortable, soft-cushioned hunting seats. Once they were comfortable, they began to fit right into the challenge and learned how to get into shooting position in "stealth mode" when a gobbler approached.

Prior to the hunts, shooting at paper targets to define shot spreads at different patterning distances proved to be very valuable; not only as a tool to get used to the shotguns but also as a great confidence builder when it came to making lethal head shots.

My one regret is that I didn't get them started earlier. Deer and squirrel dominated their first hunts but once they were introduced to turkey calling, they recognized the hunt was quite different. When you call to turkeys they sometimes answer you. Deer don't.

Jessica and Maggie are now in their early 20s and both have full-time jobs. And both also set aside vacation times for those addictive spring gobblers.

Oh yes, one more benefit: They now take me hunting with them.

BRIAN POST
Montrose, PA

Brian Post is the owner and chief guide at Full Fan Lodge in Susquehanna County. He's a member of the Endless Mountains Chapter and considers his most notable gobblers to be any he's called in for a kid or a first-timer.

Brian Post with two Camp Compass hunters

A GUIDE'S VIEW: FIRST COMMANDMENT, "DON'T MOVE!"

At Full Fan Lodge most of the guides do all the calling. Personally, I don't care if my hunter calls or not. It's up to him or her, although most newcomers want a guide to do the calling, especially the kids I host every year. It's an education for them.

However, I don't think you can scare a turkey with anything that sounds remotely like a turkey call.

Safety is my primary concern when hunting with someone I don't really know all that well. I can tell in five minutes how safe my hunter is with his gun. If he isn't safe, we'll talk about it immediately and that usually takes care of the problem.

With today's TV shows and videos, most people know what and where to shoot at the turkey. Many of them bring guns designed for turkey hunting. The equipment has come so far in the last 10-15 years, it is truly amazing.

I remember making fatal head shots on turkeys when I was a kid. My shotgun was the same one I hunted rabbits with. Back then, it was all we had.

I guide a lot of young hunters and older people who have very little in-the-field experience. I really enjoy being with a person on their first encounter and with a good, gobbling turkey. When it's all over, whether we kill a bird or not, we've had a good time.

I can say by far the number one mistake novices make is not understanding what "DON'T MOVE" means. Don't point. Don't move the gun. Don't even blink if you can help it. I would rather have a hunter in army fatigues who can sit still than a client in a $500 camo suit that can't.

I tell them all: "Turkeys make a living seeing. The trick is to not let them see you."

A sly tom knows when to gobble and when to keep its beak shut.

CHAPTER 37

FIRST HUNT

Never Thumb Your Nose at a Gobbler

My wife's eyes locked on the ridgeline 40 yards ahead, trying to make the budding scrub oak part for a look at the gobbler that rocked the Sullivan County woodlands with its incessant bellows.

He was close but as yet unseen.

I raked the wooden peg gently over the slate, hoping the wary bird wouldn't pick out the subtle hand movement and depart without us having laid eyes on him. Twenty minutes later neither its footsteps on the dry leaves or its chilling voice was heard. The sly tom had slipped away.

We'd played games with him all morning long. Even when thunderclaps that promised rain but never delivered rang out the bird would answer.

"He's hot," I'd muttered when he double- and triple-gobbled at my imitations.

"But he's not dumb," I added 90 minutes later.

I did notice a pattern, however. Whenever Betty Lou and I cautiously circled to the western edge of the ridge, he'd gobble to the east. A long circular stalk to the east and another series of yelps brought an answer from the west.

Playing the gobbler's game

Glancing at my watch, I was surprised to note that we had but 30 minutes until the opening day quest would come to a close; then 11 a.m. Perhaps there was time for one more ploy. We'd beat him at his own game.

I placed Betty Lou atop the western end of the ridge where the broad-leaved woods yielded to a dense stand of hemlocks. Fresh tracks and droppings confirmed he'd been there, at least nearby, several times during the morning game of hide-and-seek.

"You stay right there and watch the area around that stump," I hastily ordered as she nestled her back against a hemlock trunk and pulled up her mask. "And don't move."

I hustled back to the spot at which I'd been calling earlier, nestled my back against the same red maple and began a new series of yelps and cuts. But all was silent and when the watch-hands signaled that the morning's hunt was over I wandered back in the direction where I'd abruptly left Betty Lou sitting. Circling in from behind, the direction from which I'd departed minutes earlier, I saw her

The ill-fitting glove's "empty" thumb momentarily blocked the view from Betty Lou's right eye. When she closed her left eye upon aiming the gobbler "disappeared."

rushing toward me, her reddened cheeks nearly as bright as her lipstick.

"He was right there. Right where you told me. I looked away and when I looked back there was this red head. But when I went to shoot it disappeared and ......" her voice trailed into a garbled rerun of the events for which turkey hunters live. Her blood pressure and adrenalin were on high.

"He was right here," she continued, pointing to the end of the deadfall. "He came right by that log where you told me to watch."

Thumbing her nose

The log was but a mere 20 yards from her set-up spot, a perfect distance for a clean kill. As the red-headed tom strode close, pecking here and there in the forest litter, Betty Lou peered down the barrel, gently nudged the safety, closed her left eye and ..... the bird disappeared. Opening both eyes once more, it reappeared.

For a few seconds she remained perplexed, then it struck her that she was wearing two left-hand, camo gloves. On our hasty departure from the cabin that morning she'd grabbed the similar gloves, one of them mine. So she simply put the larger, ill-fitting glove on her right hand – palm-side back. But what she didn't give due consideration was that the glove's thumb didn't have her own thumb in it. When she realized that the long thumb was blocking her aim she twisted her right hand a few inches away from the side of her nose. The big gobbler caught the mini-movement, leapt into the air, set its wings and glided down into the dark hemlock bottom.

Close ….. but no longbeard this day.

Such are the memories of the ones that got away; memories that are told and retold over the years.

I knew then that I'd sold her on turkey hunting, which she'd resisted despite taking readily to hunting deer, squirrels and waterfowl for the past few years. Since that warm morning, however, she's taken 15 or so gobblers, by herself and with partners in Pennsylvania and beyond. She's just about given up on ducks and doves, opting instead for bucks, bears, pronghorns and gobblers as her most cherished pursuits – the latter urged on by a light-hearted "revenge" on the Sullivan County tom at which she literally "thumbed" her nose.

◆ ANOTHER VIEW ◆

RHONDA HENRY Friedens, PA
Rhonda Henry is the coordinator of the Whitehorse Mountain Chapter of the P-NWTF's Women in the Outdoors. She regularly visits the turkey woods with her father and husband. Rhonda's favorite gobbler county is Somerset.

WOMEN HAVE THEIR OWN TURKEY TALES TO TELL

It is 4:30 a.m. on the first day of the 2003 spring turkey season in Pennsylvania. I wake up to rain, but that doesn't make me any less excited about hunting today.

My husband Doug and I are out the door to meet my dad by 5:15 a.m. We all live on the family farm, so we discuss where we will be setting up and then it is off to the hunt.

The rain has stopped.

As we are ready to get out of the car the downpour resumes. We sit in the car and wait 10 minutes and are pleased when it slows to a drizzle. I get myself situated at my spot as, happily, the rain finally stops. Not five minutes after sitting down, a turkey flies down from the trees behind me, over my head, and into the field. "Turkey fever" begins to take hold as I watch two more dark forms 100 yards away on the other side of the field. I work my Turkey Duster Push Button call, continuing off and on for 15 minutes. But the birds go the other way.

About 10 minutes later my dad decides to change position and as he walks by we talk for a moment. He glasses the field and sees birds about 500 yards away.

We both sit down as Dad gets out his paddle box call and strikes a few notes. A pair of turkeys come running our way. When they get to 50 yards we see they're hens but that's O.K. There's nothing like "live decoys."

Out of the corner of my eye I see a turkey coming through the field and he is silent. It's a jake. When he spots the hens he runs at them, making a strange sound I can only describe as a "chicken with laryngitis."

He continues to close in, then all of a sudden takes off running out through the field toward the hens and making the same, strange sound like I have never heard; definitely not a gobble.

Before long the hens are beside me and one is eye to eye with Dad, who is behind a log. I know at this point, I cannot move because if I do the hunt is over.

As the jake comes back towards me through the field he also starts to head towards the woods right where I am sitting. At 20 yards I pull the trigger. I am thrilled as the bird flops and I realize I harvested my first turkey and I did it with my dad, the one who taught me to shoot and first took me hunting from little up. We have hunted deer together in the past, but this was our first turkey hunt together. We both shed a few happy tears that day.

A year later (2004) I bagged my second turkey, another jake, in the company of Doug. The two most special will always share those memories.

When I look back, I see that women are a definite part of keeping our hunting tradition alive. Women are now helping with habitat projects, harvesting game, buying hunting clothing and educating other women and kids.

In recent years I have met many women through the NWTF's Women in the Outdoors Program for which I am a coordinator. Many of those women were introduced to the shooting sports and have taken part in hunting or other outdoor activities.

Women are now going to hunting camp together for ladies hunting weekends or on fishing trips and canoeing outings. The sky is the limit for us. When men and women get together, it isn't only the men telling their hunting stories.

Now we have a few tales of our own to share.

WITO events include hands-on activities with a balance of
mini-seminars on a variety of topics.

Nancy Craft and son Dave

NANCY CRAFT
Hughesville, PA

Nancy Craft has been hunting turkeys with her husband for 37 years, often in the company of their infant children and grandchildren. She's president of the Muncy Creek Chapter of the P-NWTF. Nancy hunts primarily in Sullivan County.

Turkey Hunting's a True Family Affair

Gobblers are very much creatures of habit... and family tradition.

After many years of observation, I have found that gobblers tend to feed and roost in the same general area day after day. They also follow the same basic paths as they travel throughout a given area. I also learned that if a bird is missed or bumped one day, he's very likely to be in the same general area the following day. A turkey would have to be badly spooked to move to a new area.

Patience is an important asset when turkey hunting. Birds will return to the same area where one has been shot within a very short time. A few years ago my husband called in two gobblers traveling together. He shot the first bird and took it home. An hour later he returned to the same spot with a friend and successfully called in the second bird for his friend.

Turkeys are very intelligent creatures. They know where they can wander and be safe around human activity. We've had turkeys feeding in our front yard completely ignoring us sitting on our porch.

But those same turkeys have a totally different opinion of us when we enter the woods. As long as we're around our house they see us as no threat. When we enter the woods we become predators and they react as prey.

A few years ago our dog was lying in our yard and a turkey hen and several poults were nonchalantly feeding on mulberries about 20 feet from the dog. The dog and turkeys totally ignored each other.

Turkey hunting can easily become a family affair. My husband and I began taking our son turkey hunting with us when he was two years of age. We would arise before daylight, then take him, a sleeping bag, our turkey gear and head for the woods.

After setting up on a bird, we'd place our son in the sleeping bag and begin calling. We bagged several birds over the years when accompanied by our child.

Today he is an avid turkey hunter, and we've already begun taking our three grandchildren gobbler hunting with us. The tradition is being passed on and we're spending quality time with our family.

TAMMY MOWRY
Butler, PA

Tammy Mowry is the five-state regional coordinator for Women in the
Outdoors. She is also a past board member of the Pennsylvania chapter of
the NWTF. Tammy hunts Butler, Warren, McKean and Forest counties.

Women in the Outdoors: From Bird Watching to Bird Hunting

The NWTF's Women in the Outdoors (WITO) program is dedicated to providing women with educational outdoor opportunities that are both exciting and interactive. The federation launched the program to achieve its goals of teaching women about wildlife management, increasing their participation in outdoor recreation and encouraging support of our mission to conserve the wild turkey and to preserve the hunting tradition.

The first WITO day was held in 1998. That year, Pennsylvania also held its pilot event to get the program introduced to the local chapters. The first such event was held by the Moraine Chapter in June, 1999.

Since the program's beginning, more than 46,000 women nationally have participated. Since 1999, in excess of 3,400 women set aside time to take part in over 100 WITO in Pennsylvania alone.

Programs range from single-day programs to weekend getaways. Participants had their choices of seminars and hands-on activities including hiking, biking, hunting, fishing, shooting, canoeing, rock climbing, horseback riding, shotgun, handgun and rifle shooting, archery, canoeing, cooking, fly fishing, rock climbing, bird watching, caving, first aid, self protection, map and compass use and scuba diving.

The women who attend the WITO events come from all sorts of backgrounds; housewives, accountants, secretaries, lawyers, teachers, students, telecommunications experts and many more vocations.

It is always satisfying at day's end to hear women make comments like: "Had a great day;" "Good instructors, very informative;" "It is fun and everyone is great;" All workshops were excellent;" "Thanks for letting me do pistols and gun care;" "The instructors were very patient and explained everything well."

Thinking about joining?

Research shows that the following demographics among WITO members are revealing, as follows: 39 is the average age; 64% are married; 63% live in rural areas; 36% are hunters; 34% were members of WITO prior to attending an event; 88% are registered voters; 34% have never hunted, but want to learn; 73 percent have a firearm in their home; 55% of those that own a firearm use it for both recreation and protection; and shotgunning, fly fishing and archery are the top activities being tried for the first time.

For more information on Women in the Outdoors and its Pennsylvania events, call Tammy Mowry at 724-284-9201.

SECTION X
The Hunters Who Made it Possible
The NWTF and its Pennsylvania Connection

Women in the Outdoors is one of the NWTF's most successful outreach programs.

CHAPTER 38

PROFILE: PENNSYLVANIA
CHAPTER OF NWTF

22,000 Members and Never
a Dull Moment

By Don Heckman

The Pennsylvania Chapter of the National Wild Turkey Federation was organized in June 1975 by a small group of dedicated turkey hunters looking to get involved in wild turkey management and habitat.

Today the state has 72 chapters composed of more than 13,000 adults and 8,000 junior members.

The goals of three early Pennsylvania chapters centered on establishing and maintaining public interest in the wise management and conservation of the wild turkey across Pennsylvania.

Don Heckman of Camp Hill addresses the NWTF convention after receiving the organization's Distinguished Service Award.

At the very first meeting in State College, PGC Wild Turkey Biologist Jerry Wunz presented a slide program on the "Status of the Wild Turkey in Pennsylvania." He reviewed past problems and then-current programs being conducted by the PGC to help improve habitat and overall welfare of Pennsylvania's wild turkey population. Wunz and fellow biologist Arnie Hayden recommended and helped introduce Pennsylvania's first spring gobbler season in 1968.

In favor of trap-and-transfer

We also were closely involved in the pen-raised turkeys issue, eventually resulting in closure of game farms and adopting trap-and-transfer as the future of Pennsylvania's hens and gobblers. That was followed by supporting the

Biologist Jerry Wunz worked closely with the then-new Pennsylvania NWTF chapter beginning with its origin in 1975. He died in February, 2004.

controversial closure of fall turkey hunting in counties receiving relocated turkeys. Also on the growing list of items needing attention was the re-evaluation of fall season dates in northcentral Pennsylvania and taking realistic looks at winter feeding activities and tree and shrub planting. Improvements in turkey harvest counts also remained high on the chapter's priority list.

Other chapter-supported projects included regeneration of oak forests, leg banding of relocated wild turkeys, researching life spans of turkeys in the wild, proposing a wild turkey hunting license, and the establishment of Turkey Management Areas (now Wildlife Management Units) as management tools.

The reward was the wild turkey population expanding from less than 30,000 in the 1950's to over 350,000 statewide by 2003. With the new WMUs in effect, the commission revised its lengthy "Management Plan for Wild Turkeys in Pennsylvania" to achieve maximum carrying capacities throughout its vast range.

Safety first

Turkey hunter safety programs were also established. Highway billboards displayed the turkey hunter safety message "Be Sure - Identify Your Target" and turkey hunter safety posters were posted on state gamelands' bulletin boards. It was also distributed at sports shows and to participants in hunter education courses.

The Pennsylvania Chapter of the NWTF also sought and received partnership agreements with the Game Commission, Department of Conservation and Natural Resources, the federal Corp of Engineers, Allegheny National Forest and the U.S. Forest Service. The efforts quickly established communications and working agreements for habitat, safety, education and land improvement projects. Since 1985 the list has grown as has the number of Keystone State chapters, which stood at an astounding 72 in 2004.

Biological management

The Pennsylvania chapter firmly believes in wild turkey management based on biological research – including field and technical data for the successful implementation of the PGC's published management plan.

We also support funding the plan, a second spring turkey tag with a stamp for a fee and the reduction of blaze orange requirements for turkey hunters.

In 2003, state and local chapters exceeded the $3,000,000 mark for funding NWTF Super Fund projects in Pennsylvania. From 1985 to 2004 a total of 1,651 Super Fund projects have been funded and completed, raising $3,306,456. Chapter volunteers continue to set the standard that all other conservation organizations now use for funding habitat, hunter safety, land acquisition, and wildlife management.

Staying in touch

Communications are a vital link to NWTF members. The Pennsylvania Chapter publishes a quarterly, color newsletter/mini-magazine titled "Turkey Talk" for all Pennsylvania members. NWTF publishes the award-winning "Turkey Call" magazine six times a year and "The Caller Newsletter" is published quarterly for every adult member.

Many other programs and projects keep the Pennsylvania chapters busy and productive. One of my favorites is the Juniors Acquiring Knowledge Ethics and Sportsmanship – JAKES – program for children and young adults. It exposes youth age 17 and younger to hunting, fishing, conservation, and the great outdoors.

Hunting Heritage is NWTF's legacy for all hunters, established to aid volunteers and chapters to preserve the hunting and shooting traditions. The NWTF and its chapters believe such a program is vital to the success of hunting, conservation, and the shooting sports.

The Pennsylvania NWTF chapter has purchased everything from farming equipment to seeds for the development of gamelands' food plots.

For more information, log onto the Pennsylvania chapter Web site at www.panwtf.com or the national Web site at www.nwtf.org

Don Heckman is the executive officer of the Pennsylvania chapter, serving since its origin in 1985.

Jerry Zimmerman

CHAPTER 39

PENNSYLVANIA PRIDE

Driven by a Passion to Give Something Back

By Jerry Zimmerman

Pennsylvania's National Wild Turkey Federation (NWTF) volunteers are some of the most dedicated, loyal, and hard-working volunteers in America. This dedication can be attributed to a variety of things, but most of all is that these folks have a burning passion. They foster a strong desire to give something back to wildlife and the wild turkey and to protect and pass on the hunting tradition to our children, grandchildren and their children.

Pennsylvania's NWTF volunteers lead the country in the money deposited and spent through NWTF's Super Fund system. This system is the vehicle by which monies are spent throughout Pennsylvania for various projects and NWTF outreach programs. NWTF volunteers in Pennsylvania take great pride in their accomplishments to affect wildlife habitat in the state. Nearly every one of our 74 state chapters do some type of habitat work focused on the wild turkey, but it also benefits many other wildlife species.

Cooperative ventures

Our chapters work closely with the PGC land managers, the DCNR Bureau of Forestry district foresters and various other county and municipal agencies to perform projects on public lands. These projects include, but are not limited to: creation of new food plots; cutting trees and shrubs to "daylight linear" openings; top dressing existing food plots; resurrecting food plots that have been left to go fallow and bringing them back, planting trees and shrubs for winter food sources; and providing winter cover for wildlife.

NWTF local chapters have also contributed thousands of dollars to assist in the purchase of land which is then donated to the PGC or DCNR to remain forever open to hunting.

Outreach programs strong

Yet another passion of our volunteers is working with NWTF's outreach programs. The JAKES (Juniors Acquiring Knowledge, Education and Safety)

The federation's Wheelin' Sportsman program helps disabled men and women enjoy shooting, hunting and fishing.

youth program has been and continues to be the favorite of local chapters, as stated by Don Heckman in his article. Our volunteers realize that the future of hunting and conservation lies in the hands of today's youngsters. Our local chapters strive to teach the children the importance of conservation and the major role that hunting plays in conservation. The youth are taught that hunting safety is paramount when they are afield, and they are taught the importance of being knowledgeable and ethical hunters.

Yet another outreach activity growing rapidly throughout Pennsylvania is our Women in the Outdoors program. Many local chapters in Pennsylvania are hosting events that provide women the opportunity to learn more about the outdoors and hunting. Pennsylvania local chapters are among the leaders in the country in the number of events hosted in this state.

NWTF's newest outreach program is Wheelin' Sportsman, providing disabled individuals with the chance to experience the outdoors through shooting events, fishing events, and various types of hunts.

Pennsylvania's local chapters are involved with their communities, as well. A majority of the local chapters participate in Turkey Hunters Care, a program where local chapters purchase turkeys and provide them to needy families around the holiday seasons.

I started by writing that Pennsylvania's volunteers do what they do because they have a passion. To do all the things that have been described above, a person has to have desire, time, dedication and a touch of talent.

But most of all, they require the passion to want to make a positive impact on their children, grandchildren, the wildlife, and the places wild creatures live.

Jerry Zimmerman of Emmaus is one of Pennsylvania's full-time regional directors for the NWTF. He can be reached by logging onto (panwtfsrfsjz@prodigy.net). Other Pennsylvania directors are Larry Holjencin (timberline@alltel.net) and Bob Farkasovsky (wvrdnwtf@aol.com).

CHAPTER 40

NWTF A CLASS ACT AT HOME AND AFIELD

Turkeys, Hunters and the Public Reap the Benefits

Edgefield, S.C. – Tucked away in this sleepy town 25 miles northeast of Augusta, Georgia is a tribute to the nation's camouflaged cadre of wild turkey hunters, numbering more than 2.6 million and growing.

It's the home of the National Wild Turkey Federation (NWTF), an organization created in 1973 to reestablish the nation's crop of wild turkeys, which had dropped to an all-time low in 1930 and were just beginning to wing upward 43 years later. By the time the federation was formed, turkey numbers had grown to about 1.3 million thanks to the enforcement of hunting laws,

A larger-than-life bronze of a family of turkey hunters greets visitors to the South Carolina headquarters.

An animated elder tells the story of Old Mossy Toes from the porch of his Appalachian cabin.

trap-and-transfer programs and the regeneration of fields and forests once cleared for agriculture and timber interests.

Today, an estimated 6.4 million hens and gobblers live in 49 states and five Canadian provinces, a success story unmatched in conservation history. Just as astounding, perhaps, is the federation's growth to 500,000-plus dues-paying members, well above the 1,300 members carrying membership cards in 1973.

"And it's not over yet," said NWTF executive vice president and CEO Rob Keck, a former Perry County, Pennsylvania art teacher who greeted my wife and me when we visited the impressive headquarters in a recent summer. "In the last year alone we've grown by over 100,000 and our goal is to be a million-member organization in seven years."

Keck, who also hosts the NWTF's "Turkey Call" television show airing on the Outdoor Life Network, said "investing in people, not in mail order" has been the key to the federation's success. We compete with other organizations for money, yes, but we compete mostly for volunteers," Keck said, "and that's our strength."

The NWTF's staff of 215 includes workers based in the eastern South Carolina headquarters and dozens of regional representatives scattered around the nation.

Visitors welcome

Visitors to Edgefield, however, are treated to yet another aspect of the organization that may otherwise go unnoticed; the NWTF Museum. Each day a bus load or two of kids show up to learn about turkeys and the heritage that is turkey hunting. Members and non-members traveling down I-26 also drop in, as do many other travelers.

Visitors are first struck by a detailed, larger-than-life bronze in the headquarters' garden; a man, woman and boy returning from a turkey hunt, "depicting the past and the future of our hunting heritage."

"This is the only museum of its kind dedicated solely to wild turkey restoration, management and hunting," said Keck as we entered a room filled with fourth-graders captivated by a life-size, animated Cherokee Indian story-teller. The realistic mannequin lifts his arm, cocks his head, rolls his eyes and tells of the importance of turkeys to his tribe. Around the corner another animated old-timer sits in a rocker on the front porch of an Appalachian cabin, reminiscing about a long-ago turkey hunt for "Old Mossy Toes" that sticks in his memory.

Kids and adults seeing the exhibit for the first time become hypnotized by the computer-controlled figures that appear defiantly real.

Take a mini-hunt

But it's the Dave Harrelson Memorial Theater, a small, dark cornerless room with several wooden benches, in which visitors become consumed by the sights and sounds experienced by a springtime hunter in an eastern woodland; a mini-hunt, of sorts. Entering in near darkness, the doors soon close and the woods slowly awakens with the sounds of a vernal dawn. Frogs chirp, crickets call, a whippoorwill sings, barred owls laugh to one another and, finally, a gobbler breaks the stillness of morn with its limb-shaking call. Simultaneously, a start-studded sky slowly yields to dawn's early light along with the symphony of forest birds awakening.

World-class dioramas featuring each of the turkey world's six subspecies of wild turkeys, a display of box calls from legendary call makers including Neil Coss and M.L. Lynch and other displays round out the one-of-a-kind museum.

The Wild Turkey Visitor Center and Museum also includes a 100-acre Outdoor Education Center composed of nature trails, a wetland habitat and a wildlife management demonstration area.

A quality crowd

Despite the big changes in the NWTF over the years, some things remain the same. Writing in the May/June 1993 issue of Turkey Call, well-known turkey activists Wayne Bailey and Neal Weakly wrote: "Probably the single most important factor in the success of the organization was the quality of people it attracted. For reasons not fully understood, the board of directors consistently included individuals of foresight and talent who were willing to work hard, often at great personal expense, for the good of the federation."

If you find yourself in the Edgefield region on your next trip south, schedule a stop at the NWTF facility to view the displays, meet the employees and feel the pulse of this one-of-a-kind conservation organization.

For additional information, call 1-800-THE-NWTF or log on to www.nwtf.org.

ROB KECK
Edgefield, S.C.

Rob Keck is a native Pennsylvanian widely known as the CEO of the NWTF and host of its popular Turkey Call and Turkey Country TV shows. One of his favorite hunting partners is his daughter Carolyn.

PENNSYLVANIA PROUD: MEMORIES ARE MADE OF THIS

By Rob Keck

Growing up in Lancaster County during the Soil Bank era and the prime ringneck pheasant years turkeys, though low in numbers, were beginning their return. They held a special place in hunting stories told by my Pop Keck, Dad, Uncle Len, Uncle George and Uncle Bill.

Going "up north to the mountains" always held the promise of seeing wild turkeys. Sighting that first flock in 1957 with Pop and John Hoover in Lycoming County made a lasting impression, which has driven me to answer the call ever since.I fondly remember that for Christmas in 1958 all I wanted was a Stevenson's box call, which was made in Wellsboro by the late Louie Stevenson. It was a call that I saw Mr. Stevenson demonstrate on Harry Allaman's *Call of the Outdoors* show which then ran on WGAL-TV. The show, which began in 1955, was produced and hosted by Harry, a Col. Sanders look-alike. That was long before videos and The Outdoor Channel. Every hunter and angler who received it via antenna made it a point to flick on the set at noon each Sunday. The "must see" weekly show had significant influence on me.

My first turkey hunt for television took place in spring 1975 with Harry running the camera (16mm film, not video) on my good friend and calling mentor, three-time Pennsylvania state calling champion Chet Lesh of Ickesburg. On that initial show Lesh, who coached me to the state title in 1974, and I called for Sheila Link who was then president of the Outdoor Writers Association of America.

Little did I know it at the time but what a boost it was to my career. Later I had the honor of appearing on the Call of the Outdoors show with its new host Tom Fegely, after Harry passed on the hosting and filming duties in 1979.

That Perry County tie to Chet Lesh was a significant period in my life. As a schoolteacher, having the opportunity to move to the turkey calling capitol of the world, Perry County, was another dream come true.

Pennsylvania at that time was the hotbed for champion callers and some of the most unique innovations in calling devices. Examples of these new inventions included the double and multiple reed mouth calls developed by state champion caller Glenn Fleisher of Newport who was later killed in a turkey hunting accident.

Who can forget the double slate calls invented by my close mentor George "Fritz" Fleisher of Millerstown? They were made famous by his good friend across the Tuscarora Mountain, D. D. Adams of Thompsontown. I also recall the Shaker Gobbler Call developed by George Fleisher and sold to Penn's Woods Products of Delmont which was then the largest manufacturer of turkey calls in the country.

My move to Perry County was the perfect choice at the perfect time. Spring turkey hunting was still brand new to Pennsylvania hunters and hunting and calling techniques, strategies and the calls themselves were evolving year-round at a feverish pitch. As a bachelor hunter, I was privileged to be mentored by the likes of Wayne "Kochie" Kochenderfer, Fritz Fleisher, the Fleisher Family, Chet Lesh, Wayne Weibly, Dick and Richard Smith, Dale Rohm and the Rohm family including Terry, Robby and Putt, Bill Swartz and so many others. I was living a little boy's dream, going to calling contests, conducting calling seminars, hunting for Penn's Woods, playing with calls and calling nonstop.

Who could ask for more?

In 1976, Terry Rohm, Dick Smith and I traveled to Mobile, Alabama at the invitation of our good friend, Ben Lee, to compete in the World Turkey Calling Championship. We brought a new sound to this southern dominated championship and left with three of the top five trophies; my world championship, Terry as runner-up and Dick as fourth runner-up. It was the first time in almost 50 years that a Yankee won the world title, giving legitimacy to Pennsylvania callers. Those years in Perry County also took me to the conservation side of the wild turkey. While working for Penn's Woods as program director I joined Frank Piper, yet another mentor who was then president of Penn's Woods, in State College for the formation of NWTF's Pennsylvania State Chapter. Turkey project leaders, the Pennsylvania Game Commission and its turkey biologists Jerry Wunz and Arnie Hayden all had a dramatic and lasting effect on wild turkey management in Pennsylvania and, eventually, far beyond its borders. This good science helped mold my conservation principles. The big issue at that time was the need to close the Loyalsock Game Farm, which raised semi-wild turkeys to be released in marginal and submarginal habitat. It took a tremendous education and media effort to show the hunters of the Commonwealth that the game farm was little but waste of money and effort and did nothing to reestablish wild turkeys.

The author of this book, Tom Fegely, played a role in closing the game farm through his magazine and newspaper outlets and by publicizing the NWTF's Pennsylvania Chapter and its work.

I never dreamed that turkey hunting would become what it is today. But I consider myself privileged to have my roots in Pennsylvania soil and I shall always remember the many people who became a part of me – and a part of the call of the wild turkey.

SECTION XI

A Question of Hunter Ethics
How We Act When No One is Watching

*Mentors make indelible impressions on kids by their actions afield.
Here Barry Haydt and Curtis Evens of Danielsville and Ed Kauffman,
Wellsville, admire the hen the youngster shot in Carbon County.*

CHAPTER 41

MINDING OUR OUTDOOR MANNERS

Shouldering Our Individual Responsibilities

Ethics.

It's a nebulous term, of sorts, as it concerns hunting, fishing and life in general. Webster defines it as: "Moral principles or actions; standards of conduct."

I define it as how we act afield when no one's looking.

During a mid-morning lull in a turkey hunt an old-timer, veteran of more seasons than I, once defined it more in hunter's terms. While not in these exact

The cooperation of wildlife conservation officers and hunters goes a long way in identifying unethical acts and violations.

words, he described ethics as "Anything that hurts the sport, no matter if it gives someone a bad impression of us or does anything that takes away from the tradition or has a negative impact on wildlife, is not ethical."

His insightful comment was followed by a puff on the pipe that incessantly hung from his teeth and a lingering gaze into the cloudless sky, as if to allow me time to digest his thoughts.

Others have variously described ethics as "outdoor manners," "not breaking the law," "knowing your sport through first-hand experience," "having respect for the game we hunt," "hunting safely," and "not condoning hunters who poach or exceed the bag limits."

It's really all of these – and much more.

When you least expect it

My most recent "ethics encounter" occurred on the second Saturday of the 2004 gobbler season. I'd spent the morning in my 35-acre woodlot in hopes of enticing a tom or two to my decoys, but although they cooperated audibly from across a blacktop road, they made themselves scarce visually.

On the way back to my house about 10 a.m. I decided to give my stepson a call,

The author poses with his unintended "double" during the spring gobbler season.

inviting him to spend the rest of the morning on the woodline near my neighbor's greenfield, on which I had permission to hunt. I advised him to "sit where I placed the decoys above the house on the ATV trail" via my cell phone, then began my walk back to the house 150 yards away.

That's when I saw a red head poking from the long grass some 60 yards away. Then another. I stopped in my tracks, fearing I'd been busted. But as luck had it, the pair of sunlit toms hadn't seen me in the shadowed woods.

I dropped to my knees, crawled to a tree 15 yards away and set up. A minute later, on my initial soft yelps, both birds sounded off. Ditto for the second call. I put my box call down and shifted the gun on my knee.

Within two minutes I saw them walking up the path, which would lead them to within 15 yards of me. They were still 40 yards away when the head of the lead bird stiffened and stretched. I thought I'd been busted but then I realized he'd seen the decoys 50 yards up the winding trail.

The sighting of the fake hen and jake urged them to change their direction. Seconds later the lead bird reappeared at about 25 yards, having passed through a cluster of barberry which grows in abundance. The second tom was nowhere to be seen. When bird number one again stretched its now-white head to peer at the decoys I fired.

The bird tumbled, flapping its wings wildly as turkeys will do when hit with a load of No. 6 magnum pellets.

I arose, only to note that more than one gobbler was flopping about.

I'd shot my first-ever double; an accomplishment that would have been legal in many of the states I've hunted.

But I had no intention of doing it. Blame the thick and prickly barberry. No, blame me.

Upon later analysis, I realized the second bird had probably drawn alongside the first with its head down, but its presence was blocked by the waist-high barberry.

Leaving the birds where they finally stopped thrashing, I immediately walked to the house and called Northampton County WCO Brad Kreider, to whom I later wrote a $20 "restitution fee" check and relinquished one of the birds. One tom weighed 21 pounds, the other 20. Both sported 10-inch beards and spurs exceeding an inch.

I began to feel a bit better about the forgivable mistake in judgment after talking with Kreider.

What bothered me more, perhaps, was the response from several people who heard my story.

"Why didn't you just eat the damn thing?" one asked.

"You should have put your wife's tag on it," suggested another.

"No one would ever have known as long as you didn't brag about it," offered another.

"I'd have known," I assured him. "And I'd have had to lie to my friends and kids each time we told our tales about the spring hunt.

Ignoring safety zone and no trespassing signs is both a legal and ethical violation.

This is not being written as a catharsis, rather as a somewhat disturbing footnote to the event, which I'll surely share with other hunters and friends from time to time.

It's a matter of ethics.

But outdoor ethics isn't only about righting wrongs.

Setting our own codes

What better time to give some serious thought to the problems caused by hunters who trespass, litter, stalk and shoot roosted birds and waste game or those who commit the more serious transgressions of poaching, exceeding the bag limit or flagrantly breaking the law? As in all moral judgments, setting up one's own code of ethics is a personal matter. I can't always abide by yours. You can't unquestionably honor mine. Not entirely, anyway.

In competitive sports there's a referee or umpire close at hand to penalize the person who breaks the rules. In hunting, no such judge is on hand. Perhaps that's what makes hunting and fishing alone, one on one as many of us prefer to do it, the purest of pursuits.

I recall some experiences during my rookie days afield when illegal or "unethical" acts that I observed went without comment or response. My encounters with both non-hunters and avid hunters over 50 years I've been hunting aren't unique.

The decision of what to do about ethical predicaments lies within one's personal code of living and the situation at hand. They're not as cut and dry as was my springtime dilemma. Whether to report a violator to authorities, give him a personal reprimand and refuse to again hunt with him, or ignore his indiscretions are the basic choices to which such quandaries are distilled.

No matter which is chosen, one's own ethics are put to the test.

I'm most concerned about the 80 per cent or so of the general public who know little, and couldn't care less, about hunting. These are the suburban property owners whose homes are sprinkled with shot on the small game opener. They're the farmers who find their fields blocked by vehicles, gates left open, signs shot up, wire fences ripped from their staples and standing corn knocked down by hunters. Some don't even show the courtesy of rapping on the door and asking permission to hunt.

Set the example

Number one in our thoughts as experienced hunters should be setting example. Kids, especially, are heavily influenced by the actions of adults. If Uncle Clod trespasses, shoots and extra mallard, leaves a gray squirrel to rot in the bushes, tosses a can in the leaves, trespasses on private property or shows blatant disregard for the rights of others afield, then the mold will have been set for an impressionable youngster.

How many times have you, unwittingly perhaps, set a thought train for a newcomer to the sport – good or bad?

Most hunters would probably be reluctant to stand up in court to testify against a violator, especially if it's a person that may be a resident of the same town or a member of the same hunting camp. But in recent years I've heard from a surprising number of people who hunt, voicing concern over transgressions their fellow hunters, mostly friends or at least casual acquaintances, have performed.

It's enlightening and satisfying to see the growing intolerance among sportsmen for those who flagrantly bend or break the rules.

I like the statement by the late Pennsylvanian Seth Gordon, former conservation director for the Izaak Walton League, who said: "Your outdoor manners tell the world what you are when you're at home."

I also respect the thoughts of my good friend Jim Casada of South Carolina in his book "Innovative Turkey Hunting" (Krause Publications; 2000). He writes: "Ethics remains an elusive concept. Yet, deep down, every hunter has gut feelings about what's right and wrong. Be true to yourself, keep the hunter's faith, and let safety and ethics be a part of the mental gear that accompanies you on every hunt."

We, alone, carry the responsibilities for the survival of hunting as a proud and respected activity. Indeed, for some of us, a lifestyle. And as sportsmen, in the truest sense of the word, each of us must vow to educate, inform, share and uphold the highest of standards each time we tread in field and forest.

When we hunt, we shoulder more than the weight of a rifle, shotgun or bow.

GREGG RINKUS
Franklin, PA

Gregg Rinkus is a founding member of the Wapiti Roost Chapter. He hunts Venango, Mercer, Clarion, Forest and Indiana counties. His most notable tom is the 20-pound gobbler he called in for his 13-year old son on the 2004 opener.

LAWS, ETHICS CODE MUST GUIDE TURKEY HUNTERS

As responsible sportsmen, each of us is guided by legal expectations and our own personal code of ethics.

I have encountered this situation many times in my turkey hunting career. For example, I do not consider belly-crawling to get into a calling position as "stalking" anymore than I do following scratchings in the fall to locate and scatter a flock of birds.

Because safety is paramount in spring gobbler season, if I were to belly-crawl to the edge of the field I would definitely wear a fluorescent orange hat. No turkey – regardless of beard length or sharpness of its spurs is worth putting my personal safety at risk.

Another issue in this scenario is the term "ambush." Many years ago, I decided that if I can't call in a spring gobbler I will not lie in wait for one to get within shotgun range. Ambushing is not illegal; it's simply a hunting technique I choose not to use.

During my lifetime of outdoor ramblings–particularly as a wildlife photographer– I have belly-crawled more than a buck private at boot camp. When it comes to hunting, though, the pursuit of a wild turkey must conform to my personal ideal of fair chase. For me, ambushing a gobbler is not fair chase.

This scenario will undoubtedly dictate the use of enormous patience. If I can see the birds and monitor their activities, I'd probably set up, watch the goings-on and simply wait to see what develops. My best chance to bag this bird would be to call the hens toward my set-up location. Since I also don't use turkey decoys, what's better to lure in a wise old gobbler than a real hen turkey? If I can entice the hens to come to me, the gobbler will typically follow and possibly offer me a good shot.

Each situation is different and dynamic. I have to be flexible enough to quickly change my tactics as the scenario unfolds. And, in most cases, this boils down to knowing what to do and when to do it.

That's what 35-years of turkey hunting experience provides including, perhaps, the biggest variable of them all–Lady Luck!

RICK GRGURICH
Allentown, PA

Dr. Rick Grgurich is a veterinarian and president of the P-NWTF's Lehigh Valley Chapter. His most notable turkey held a 12-1/2 inch beard. He does his gobbler hunting in Lehigh County.

WHAT'S ETHICAL AFIELD?
WHAT'S NOT?

One of my earliest turkey hunting experiences involved quietly exiting my truck in the dark on a solo hunt, donning my gear, and trying a locator call. It was answered immediately by a gobbler on roost – almost directly overhead. He was so close that a silhouetted beard could be seen in the available moonlight.

Now what?

Wait until legal shooting time and take him off the roost?

This, of course, is not considered ethical; no different to me than blasting an unflushed ringneck a dozen feet away.

Ethics – how we conduct ourselves in the outdoors whether anyone else is around or not – is one of the factors that may help to shape the future (or demise) of hunting. Additionally, some of my guidelines of ethics blend right into the laws of safety.

Here's my view on the question of "What is ethics?"

•   Make sure, when hunting private property, that permission from the landowner has been obtained.

•   Bring no one else along unless he/she has also been given permission.

•   Sharing your harvest, or a holiday gift or card for the landowner, says thank you and may help to secure a place for you to hunt in the future. You might also consider offering any assistance to help the landowner, such as hauling hay, painting a fence or helping post his/her property.

•   Be aware of livestock, buildings, equipment and other people on the land.

•   Respect Safety Zones.

•   Treat all land you hunt on with respect and do not litter. Pick up your spent shells.

•   Know in advance who else might be hunting the property and where they will be.

•   Do not stalk a turkey or turkey sound. Not only is this dangerous and illegal, but you may ruin another hunter's setup.

•   Be aware that judging distances and taking unwarranted shots at turkeys far out of range is another ethical sin, resulting in wounded prey. Practice distance judging in both woods and fields so you know the difference between 20 and 40 yard shots when the moment of truth arrives. Don't be too hasty to squeeze the trigger when turkey fever sets in.

• If you come upon another hunter on a set-up, turn around and leave after you make sure he knows you're there.

• Always avoid confrontations in the woods. Settle matters later in a safe, calm and adult manner. No one needs a hunt ruined by an argument.

• When you "rejoin civilization" after your hunt, (for example, going to breakfast or lunch) be aware of the image you project, your overall appearance, what you say and how you behave. This can reflect positively or negatively on all hunters.

• Finally, do not shoot birds off their roosts. Learn to respect, appreciate and enjoy the privilege of being an ethical and respected hunter.

◆ ANOTHER VIEW ◆

MARY JO CASALENA
Bedford, PA

Mary Jo Casalena has been the project leader for the PGC's wild turkey program since 1998. When not studying turkeys, she may be found hunting them in Bedford and Washington counties.

TURKEYS IN 2015:
CRYSTAL-BALLING WHAT'S TO COME

Looking ahead 10 years, I foresee higher turkey populations in many parts of Pennsylvania, especially in the northcentral and southeastern regions of the state. We currently are revising the "Management Plan for Wild Turkeys in Pennsylvania," which specifies goals, objectives and strategies for managing wild turkeys on a 5-year basis.

The initial plan was developed in 1999. The overall management goal will remain the same, but the specific objectives will be determined via public input and involvement. This plan will be a cooperative effort between the PGC and the public. Our management goal is to maintain and enhance wild turkey populations in all suitable habitats throughout Pennsylvania for hunting and viewing by current and future generations.

Some changes I foresee within the next 5-10 years include more turkey hunting opportunities during the spring season, with a second spring gobbler and, possibly, all-day spring hunting. These changes would increase hunting opportunities for youth and adult hunters without impacting the wild turkey population.

Our spring season is timed to begin after the peak of breeding. The spring hunt opens after the majority of hens have begun incubating. Therefore, breeding disturbance and illegal shooting of hens are minimized. Thus, expanding spring hunting opportunities is likely to have less impact on the turkey population than expanding the fall either-sex season.

Also, even though Pennsylvania is traditionally a fall turkey-hunting state, spring turkey hunting has become more popular since 2000. We anticipate this trend will continue.

Turkeys have expanded their range into more residential areas and with the expansion comes the possibility of conflicts with humans. Our goal is to maintain healthy wild turkey populations balanced with human populations in order to minimize conflicts.

Yes, we still have plenty to learn about our wild turkey numbers and locations in Pennsylvania. Our current turkey population model only provides a general population estimate.

The PGC is currently participating in a multi-state wild turkey population modeling study which will help states determine data needs for more accurate population estimates.

I know it's a long way off and crystal-balling is not a true science. By 2015, however, I expect a more accurate population model to better assess harvest options and better justify harvest policies for our valuable wild turkey resource.

The future looks bright.

DEADEYE DESIGN DAD

Robert G. Visgaitis, 72, of Hazleton, PA holds the hen he shot in Luzerne County on Nov. 5, 2004, two day's before this book's printing deadline.

MEMORIES OF DAD

Homer Fegely, the author's father, poses proudly with his first – and last – gobbler.

DREAMING OF BIG TOMS

When you abide by the turkey hunter's unwritten law of "don't move" a curious gobbler may wander well within shotgun range, as in this set-up shot of a sleeping hunter.

Morning Encounter

The author had his camera and flash with him before dawn on a recent spring season opener when he nearly stumbled across this newborn fawn in a Carbon County field.

Last Laugh

Joe Sears of New York reveals his secret method for calling in really big gobblers.

NWTF CHAPTERS

Local Chapters Of The Pennsylvania Chapter Of NWTF

Adams County Longbeards
Allegheny Mountain
Allegheny Plateau
Allegheny Sultans
Allegheny Valley
Anthracite Longspurs
Armenia Mountain Spurs
Arnie Hayden Memorial
Bald Eagle Longbeards
Beaver Valley Longbeards
Ben Stimaker Memorial
Blue Mountain
Cascade Thunderin' Toms
Chesquehanna Spurs
Colmont Gobblers
Delaware County Longbeards
Endless Mountains
Foothills Spurs
Fort Chambers
Friendship Hills Spurs
Gobblers Knob'
Honey Hole Longbeards
Juniata Gobblers
Kinzua Allegheny Longbeards
Kinzua Valley
Kit-Han-Ne

Lake Marburg
Lake Region Longbeards
Lakefront Gobblers
Lakeland Longbeards
Lehigh Valley Longbeards
Lenni-Lenape
Lower Bucks Longbeards
Lackawanna Longspurs
Valley Longbeards
Mason-Dixon
Michaux-Yellow Breeches
Mill Creek
Mon Valley Longbeards
Moraine
Moshannon
Muncy Creek
Nittany Valley Longbeards
Northern Counties Full Fan
Northwest Thunderin' Toms
Pennsylvania Local
Peters Creek Trail
Philadelphia County Fightin'
Spurs
Pocono Mountains
Red Rock
Red Rose

Roger Latham
Schuylkill Spurs
Shade Mountain
Shenango Valley Beards &
Spurs
Shermans Valley Strutters
Southeast Silver Spurs
Stony Valley
Sun Area
Susquehanna
Susquehanna Longbeards
Tamarack Turkey Talkers
Ten Mile Valley
Terrace Mountain
The Laurels Longbeards
Tuscarora Longbeards
Walking Purchase
Wapiti Roost
Warrior Trail
Whitehorse Mountain
Longbeards
Wilhelm
Wilson F. Moore
Wolf Creek Longbeards
Yellow Creek

PHOTO CREDITS

All photographs by the author unless otherwise credited. Photos used in Another View are courtesy of the subjects.

Betty Lou Fegely: 31 (top), 36, 40, 44, 46, 49 (bottom), 60 (bottom), 61, 67, 75 (top), 76, 77, 78, 86 (top), 95, 99, 110, 128 (top), 130 (bottom), 151, 154, 162, 165 (bottom), 170, 182, 197, 204, 222, 240, 244, 246, 254 (bottom), 256, 274, 281, 282, 290, 302, 304, back cover.

Pennsylvania Game Commission: 2, 4, 5, 6, 7, 8 (bottom), 11, 12, 13, 19, 21, 25, 39, 41, 81, 124, 165 (top).

National Wild Turkey Federation: Front cover, 8 (top), 16, 28, 29 (top), 30 (top), 58, 100, 187, 214, 280.

Tim Flanigan: 26, 296. Jack Paluh: 3. Pennsylvania Federation of Sportsmen's Clubs: 257. Courtesy of Evelyn Wunz: 15, 276. Courtesy Helen Hayden: 10. Jerry Zimmerman: 278. Mike Watson: iii, 121, 177, 233. Remington Arms Co.: 42 (top).

Roger Hayslip: 14. Tru-Glo Sights: 43. Ken Merkel: 48. Greg Caldwell: 49 (top), 209. ASAT Camo: 57. Georgia Boots: 60 (top). Frogg Toggs: 62 (top). Cabela's: 62 (bottom), 235. Primos Game Calls: 74. Glen Lindaman: 94. Walker's Game Ear: 183. Shirley Grenoble: 231 (bottom), River Valley Game Calls: 252.

About The Author

Tom Fegely has been pursuing turkeys for more than 30 years. He's broadast his calls in 25 states and has taken all four subspecies of gobblers.

Tom served 25 years as the outdoors editor of The (Allentown) Morning Call, hosted Call of the Outdoors on WGAL-TV for eight years and currently appears on the masthead of North American Hunter, Buckmasters and Whitetail News. His articles also appear in Turkey Call and other outdoors publications. He wrote the Woods-Wise column for Field & Stream and has appeared in more than 100 publications and produced and wrote seven videos.

This is his ninth book, following his acclaimed "A Guide to Hunting Pennsylvania Whitetails" published in 1994 and reissued in 2000.

Tom earned his Bachelor of Science degree from Lock Haven University in 1963 and received his Master's Degree from Kutztown University in 1969, both with majors in biology and education. After 14 years as an ecology teacher he began writing and photographing full time.

In addition to more than 100 writing and photography awards, he is the only writer to have received the Pennsylvania Associated Press Managing Editors Award three times. He was inducted into the Pennsylvania Turkey Hunter's Hall of Fame in 1991 and received the Communicator of the Year Award from the National Wild Turkey Federation in 2002.

Tom lives on 35 wooded acres in Northampton County with his wife Betty Lou, also a writer and photographer, and his chocolate lab Max, who's never written much of anything.

They have seven grandchildren, five of which show promise for becoming hunters.

It's never too early to buy a kid a turkey call. Here, the author and his grandson Colton Fegely practice turkey calling in the fall woods.